11 Chemistry of Plant Protection

11 Chemistry of Plant Protection

Analysis of Pesticides in Ground and Surface Water I

Progress in Basic Multi-Residue Methods

Volume Editor: H.-J. Stan

With contributions by
M. Åkerblom, H. Egli, P. Günther, T. Heberer,
P. T. Holland, R. L. Jones, W. Pestemer,
K. Rubach, H.-J. Stan,

With 71 Figures and 23 Tables

Springer

Volume Editor:

Prof. Dr. *H.-J. Stan*
Technische Universität Berlin
Gustav-Meyer-Allee 25
D-13355 Berlin, Germany

This series continues the handbook
"Chemie der Pflanzenschutz- und Schädlingsbekämpfungsmittel"
edited by R. Wegler

ISBN 3-540-58794-2 Springer-Verlag Berlin Heidelberg NewYork
ISBN 0-387-58794-2 Springer-Verlag NewYork Berlin Heidelberg

CIP-Code applied for

Printed in Germany

Typesetting: Macmillan India Ltd., Bangalore-25
SPIN: 10077334 52/3020 - 5 4 3 2 1 0 - Printed on acid-free paper

Managing Editors

Editor-in-Chief
Dir. u. Prof. Dr. W. Ebing i.R.
Trautenaustraße 8
D-10717 Berlin, Germany

Prof. Dr. H. Börner
Institut für Phytopathologie der Universität Kiel
Hermann-Rodewald-Str. 9
D-24118 Kiel, Germany

Prof. Dr. D. Martin
Universität Potsdam
FG Naturstoffe
Stahnsdorfer Damm 81
D-14532 Kleinmachnow, Germany

Dr. V. Sjut
Hoechst Schering AgrEvo GmbH
Postfach 27 06 54
D-13476 Berlin, Germany

Prof. Dr. H.-J. Stan
Technische Universität Berlin
Gustav-Meyer-Allee 25
D-13355 Berlin, Germany

Prof. Dr. J. Stetter
ZF - FWI, Bayer AG, Q 18
D-51368 Leverkusen, Bayerwerk, Germany

Editorial Board

Editorial

Last year, we issued a book within this series (Vol. 9) comprising several scientific aspects of the occurence and behaviour of pesticides in ground and surface waters - in part after their passage through soil layers. It is almost self-evident that this very current topic in ecological concern should be supported by a publication about the methods needed to control the fate of the pesticide residues within these areas.

Undertaking this task signified a push into new experimental dimensions for the analytical chemist, for less than 0.01 mg active ingredient per litre water has to be not only identified but also determined reproducibly.

Because of the enhanced experimental difficulties incurred by operating within this range of extremely low concentrations and in order to secure the results obtained, several different methodologies have to be used and must be further developed so that they can be usefully applied to the special problems and matrices. Furthermore, some additional methods have to be worked out to confirm the results obtained by the main procedure in order to give clear evidence to questions that may arise.

This, in consequence, led to the treatment of a relatively high number of aspects in respective chapters which - in turn - forced the editor to split them into two books of the same volume "Analysis of Pesticides in Ground and Surface Water - Modern Techniques for Multiple Residues". Part I of this volume mainly explains and describes progress made up to now in technologies and methodologies of the basic means and measures to be applied in all cases of trace organic chemical analysis in water placed under the sub-title "Progress in Basic Multiresidue Methods". Part II "Latest Developments and State-of-the-Art of Multiple Residue Methods" includes preferably attempts to apply highly sophisticated instrumentation for solving open problems in this field.

I would like to thank all the authors and the managing editor for their efforts. I hope the readers - analytical experts and other interested scientists as well as post graduate students - will be able to profit from this volume which is eminently suitable as a handbook on their desks or their laboratory benches.

Berlin, March 1995

W. Ebing
Editor-in-Chief

Table of Contents

Table of Contents of Volume 12

Introduction

Contamination of groundwater by pesticides has become a matter of public concern in recent years. Clean water is probably mankinds most valuable resource. Any sound research on leaching of pesticides into the water system must be based on reliable analytical data. This includes the unequivocal identification of pesticides at trace level concentrations and a realistic estimate of their quantity. Reliable analytical data are also the prerequisite for decision making in environmental protection and pollution control.

In the two volumes 11 and 12 of our series the authors describe the most important analytical procedures in the detection and determination of pesticides (active ingredients) in water and the sophisticated instrumentation needed for multiresidue analysis. Methods for the determination of individual pesticides are not within the scope of this volume.

The first part is structured to follow the procedures of multiresidue analysis and thus begins with a general description of sampling, extraction and clean up and deals with basic multiresidue methods.

With more than three decades experience, gas chromatography in its most modern form with the extreme high separation power of capillary columns still plays the dominant role in pesticide residue analysis

This is also due to the availability of sensitive selective detectors, such as electron capture, thermionic nitrogen phosphorus and flame photometric detectors and in particular the ease of coupling to mass spectrometry. The latter still is the most important detection technique and indispensable for the reliable identification of any contaminant at trace concentration levels. Therefore, a brief review of mass spectrometry as a detection technique is included to survey the fundamental principles and to emphasize what makes this technique the unrivaled detection method in pesticide residue analysis.

Simple and rapid screening procedures suitable for analyzing a very large number of samples are based on biochemical methods and bioassays. They contribute to the cost-effective monitoring of water samples thus making the analyses economically feasable.

Measurements within this extreme low concentration range involve the danger of obtaining erroneous results, qualitatively as well as quantitatively. It is highly recommended that analysts study the chapter by H. Egli dealing with data quality assurance.

In the second part, the state-of-the art and latest developments of multiresidue analysis are described. Two chapters deal with the automation of pesticide residue analysis in surface and drinking water by coupling of solid phase extraction on-line to high performance liquid chromatography or gas chromatography is to be considered as a major breakthrough in the field of pollution control.

A new detection method which is based on atomic emission spectrometry allows the specific detection of elemental composition in gas chromatographic eluents and opens a new dimension in screening for chlorine, phosphorus, sulfur, nitrogen or any other heteroatom containing pesticides or contaminants. A brief review of this technique and instrumentation is presented, additionally the application in a multiresidue method is described.

Mass spectrometry can today also easily be combined with high performance liquid chromatography of which the most appropriate coupling and ionization techniques are described. Additionally, a recently developed multiresidue method based on high performance liquid chromatography/mass spectrometry with thermospray is described. This method offers possibilities to analyze thermolabile pesticides previously not detectable.

Even "good old" thin-layer chromatography is going to play an important role within the choice of methods in the future. With the invention of the automated multiple development devices the technique has been enhanced in its flexibility, separation efficiency and also sensitivity to the magic 100 ng per liter concentration level.

Sampling, Sample Preparation and Preservation

Russell L. Jones

Rhône-Poulenc Company, P.O. Box 12014, 2 T.W. Alexander Drive, Research Triangle Park, NC 27709 USA

Proper collection and handling of water samples is an essential part of producing meaningful analytical results, especially at the trace levels now reported. Aspects which must be considered include sample size and container, sample labels, sampling technique, sample preservation from time of collection until receipt by the laboratory, transport and storage of samples, and recordkeeping. The characteristics of the agricultural chemical to be analyzed, study objectives and design, and the nature of the water to be sampled are important factors in determining how each of this aspects should be addressed. For many agricultural chemicals, requirements for sample containers and sampling techniques can be simplified compared to those often used in other types of groundwater monitoring programs. Regardless of the specific procedures utilized, the challenge of collecting water samples in an environment where dust and surface soils may contain residues three to four orders of magnitude higher than the sensitivity of the analytical method dictates that samples always be collected by conscientious and trained personnel. Care must always be taken to avoid contamination by attention to cleanliness of the sampler, sample bottles, sampling equipment, and storage containers. Samples (or empty sample bottles) should not be transported in vehicles also used to transport agricultural chemicals or stored in containers or freezers previously containing agricultural chemicals. Sample labels and study records must be adequate to provide necessary information and fulfil any applicable regulatory requirements.

1 Introduction

The presence of agricultural chemicals in water has been a subject of study for the past thirty years. Early research focussed on the appearance of chlorinated hydrocarbon insecticides in surface water [1] while more recently less persistent herbicides have been detected [2]. Since 1979 agricultural chemicals have been found in groundwater [3, 4]. Much progress has been made in understanding the environmental fate processes resulting in potential movement of agricultural chemicals to surface and groundwater, so that effective management practices can be developed to protect drinking water supplies from point source and non-point source residues [5, 6] A necessity for conducting environmental fate research programs is the ability to collect and analyze large numbers of soil and water samples. For example, a typical monitoring program for a specific agricultural chemical may necessitate the collection and analysis of several thousand water samples for both parent and significant metabolites at levels of 0.1 $\mu g\,l^{-1}$.

Although the remaining chapters of this book deal primarily with techniques for analyzing agricultural chemicals in water samples, proper collection and handling of these samples is an essential part of producing meaningful analyses, especially at the trace levels now reported. The purpose of this article is to provide information on the collection of water samples and the preservation of these samples until they reach the analytical laboratory. Design of research studies for measuring agricultural chemicals in groundwater and surface water (and the associated installation of monitoring wells, flumes, and other sampling devices) is beyond the scope of this paper but is discussed elsewhere [7–9]. Sediment and soil sampling are also beyond the scope of this paper. Although references to the literature have been included where appropriate, the intention of the author has been to provide more practical details on sampling than are generally available in the published literature.

2 Sample Containers

The most appropriate container for collecting water samples is dependent on the properties of the agricultural chemical being studied as well as the volume required for analysis (which may be a function of the sensitivity needed to meet the objectives of the study). Each sample also must be permanently and clearly labeled so that it can be distinguished from other samples. In order to minimize possibility of contamination from previous use, all samples should be collected in new containers whenever possible.

2.1 Volume

The size of the water sample collected depends mainly on the amount of water needed to perform an analysis. Unless the sample volume requirements are relatively large (greater than about 0.5 l), sufficient volume is usually collected to permit a second analysis. In some circumstances the limiting factor is the amount of water that can be collected. In these instances, this amount is sent to the laboratory. However, if an analysis is possible with the reduced amount, often the sensitivity is reduced. Especially when a large number of samples are being collected, keeping the sample volume down reduces the effort in handling the samples, and reduces the cost of ice or other means of cooling the samples, sample and shipping containers, and shipping and storage costs. The size of standard sample bottles also tends to influence the volume of sample collected. In studies involving the author most samples have been collected in 0.1, 0.5, or 1 l containers.

2.2 Materials

Traditionally, most sampling containers have been made of glass. Adsorption to glass is not significant for many compounds and other organic compounds are not introduced into the sample. However, glass has the disadvantage of being relatively heavy and easily broken, especially during freezing or shipment. Breakage also seems to increase with increasing container size. Therefore, whenever, possible the author's preference is to use a plastic bottle (for example, polyethylene, polypropylene, or polycarbonate). Plastic bottles are suitable for many agricultural chemicals that do not bind strongly to soil although laboratory tests should be performed to demonstrate their suitability. These tests should demonstrate that adsorption or degradation is not significant and that no analytical interferences are introduced into the sample. For some compounds, adsorption to glass may be significant, requiring the use of other types of containers. Plastic coated glass bottles may be used to reduce breakage during shipment, although such coatings are ineffective against breakage encountered during freezing. If plastic coated bottles break, the sample may be retained, but potential interferences may be introduced into the sample.

2.3 Labels

Three approaches are commonly used to identify sample containers: writing on the container with a permanent marker, attaching adhesive labels, and wiring tags onto the container. Regardless of the approach, the ink used must be waterproof and legible after being placed in water or stored in a freezer. The author's preference is to use a pre-printed adhesive label (with an appropriate adhesive and ink) since wire tags can be more easily separated from the sample container (also cardboard tags are easily torn when wet) and writing by hand on a sample container with a marker is often time consuming and sometimes difficult to read.

The information on the label must be adequate to identify where and when the sample was collected. Since samples from different sample intervals or different studies may be stored together in the laboratory, the label must not only contain enough information to distinguish between samples within a specific sampling interval, but also between samples in different intervals or studies. When conducting studies under Good Laboratory Practices (GLP) the labels should at a minimum contain the study number and a unique sample number. Additional information may be helpful in sample identification during sampling and in the laboratory. In research studies where sample locations have been determined prior to arriving at the study site, the author's practice is to usually include the company name, study number, trial location, sample location, sample date or interval, and unique sample number on each pre-printed label.

2.4 Cartridges

In some studies, immediately after collection, the sample water is passed through cartridges filled with a solid adsorbent to remove the agricultural chemical [10]. The cartridge is then shipped back to the laboratory for analysis. Advantages of this procedure include reduced shipping costs, often reduced analytical costs in the laboratory since an extraction and/or concentration step is usually the first step in the analysis, and sometimes greater stability of an agricultural chemical on the adsorbent than in water. Disadvantages include that somewhat more effort may be required in the field during sample collection (for example, plugging of cartridges is a problem under some circumstances) and that the procedure is not applicable to all materials. When using this technique, the amount of water passed through the filter must be measured to provide an accurate estimate of the concentration in the water. Validation of methods using this technique must show quantitative removal by the adsorbent and acceptable stability of the agricultural chemical in the cartridge.

3 Sampling Techniques

Numerous techniques have been used to collect water samples to be analyzed for agricultural chemicals. Typically, different procedures are used for monitoring wells, potable wells, and surface water samples. Much of the information on sampling monitoring wells has been adapted from Ref. [9].

3.1 Monitoring Wells

Collecting a water sample from a monitoring well (Fig. 1) is usually a multi-step process. First the water table elevation is measured and then standing water is purged from the well. During the purging process (or sometimes before), the water temperature is usually measured. Measurements of pH and conductivity may also be made at this time or an additional sample may be collected for performing such measurements later during the day (since pH and conductivity meters often need to be calibrated or at least checked against a reference standard, performing such measurements in batches may be more efficient). Next, the bottle is triple rinsed and then the sample is collected. Finally, all sampling equipment is thoroughly cleaned to avoid cross-contamination of later samples.

Sometimes, the recharge into monitoring wells is slow enough so that all of the standing water is evacuated before the desired amount of water is purged. If this occurs, the author's practice is to continue pumping for an additional five minutes and then begin collecting the sample. If the recharge is extremely slow,

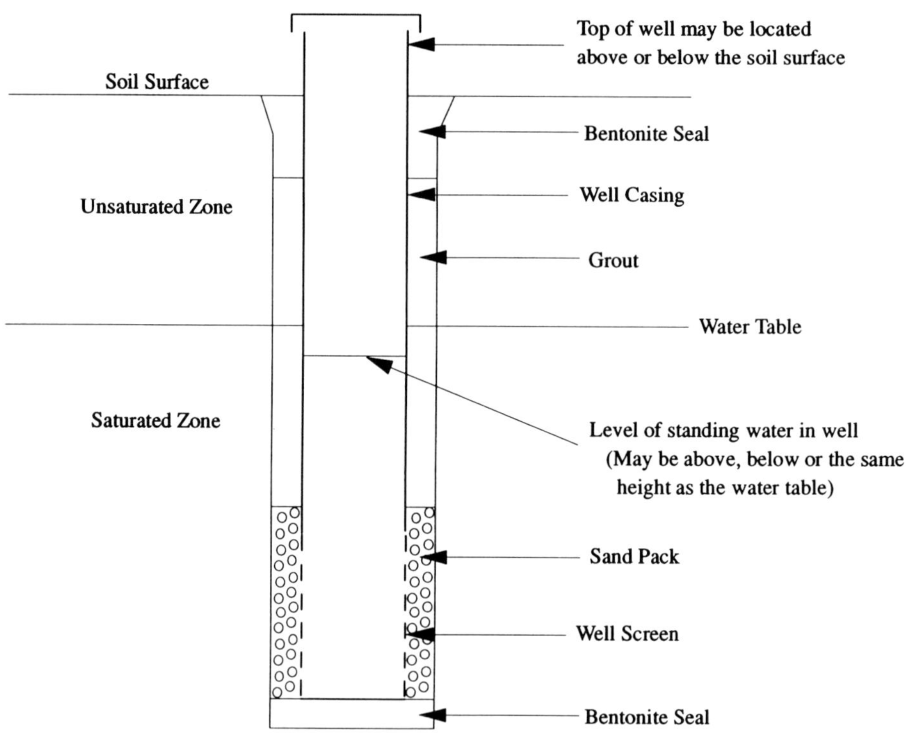

Fig. 1. Diagram of a monitoring well. The sand pack and seal below the water table will not be present for wells installed using the manual procedures described in Ref. [9]

the pump may be turned off for a few minutes until an adequate amount of water enters the well.

Two criteria are commonly used to determine the amount of water to be evacuated prior to sample collection. One is to purge until the pH and conductivity remain constant, while the other criteria is to evacuate a specified number of well volumes. Although usually 3 to 6 well volumes are typically evacuated using the latter criteria, relatively recent research [11] indicates even less purging may be adequate if the intake of the evacuation device is placed near the bottom of the well screen. The first criterion is probably the most appropriate when a representative sample of an entire section of an aquifer is desired. However, in many studies with agricultural chemicals, the constant well volume approach is preferable because the concentration of residues at a specific depth and location is desired. In the latter case, excessive pumping may result in the sample containing water which was located at a different depth than the well screen prior to the onset of pumping. Excessive pumping may also draw agricultural chemicals deeper in to the saturated zone.

A variety of devices can be used to purge water and collect samples from a monitoring well. Some of the most commonly used are bailers, pumps which are inserted into the well (such as submersible or bladder pumps), and pumps which are located above the ground surface (such as peristaltic pumps). The choice of the purging and sampling device depends on the properties of the agricultural chemical being analyzed and site characteristics. Devices which may not be acceptable for general monitoring applications such as those involving landfills may be acceptable for many agricultural chemical studies. Most agricultural chemicals for which field research is being conducted to address groundwater concerns are usually relatively non-volatile and at least somewhat water soluble. Prior to the selection of a sampling technique or equipment, laboratory studies need to be conducted with all materials such as tubing and pumps or bailers that will come in contact with the samples to demonstrate that no significant loss of the agricultural chemical occurs or that no compounds which will interfere with the analysis are introduced.

Sample collection using pumps inserted into the bottom of a well (Fig. 2) is probably the best approach when samples are being analyzed for volatile compounds although inertial pumps (Fig. 3) have also been shown to perform satisfactorily [12]. Advantages of these approaches include the ability to collect samples regardless of the depth of the water table and keeping the pressure in the sampling equipment above atmospheric pressure. Because loss of volatile compounds during sampling is minimized by the approach, the use of submersible or bladder pumps are often preferred by regulatory agencies for all groundwater sampling. One disadvantage of submersible or bladder pumps is that the well casing must usually be at least 5.1 cm in diameter and often twice as large for many submersible pumps. More water must be purged with larger diameter wells, increasing the possibility for drawing agricultural chemicals deeper into the saturated zone. Material and installation costs also increase with increasing well diameter.

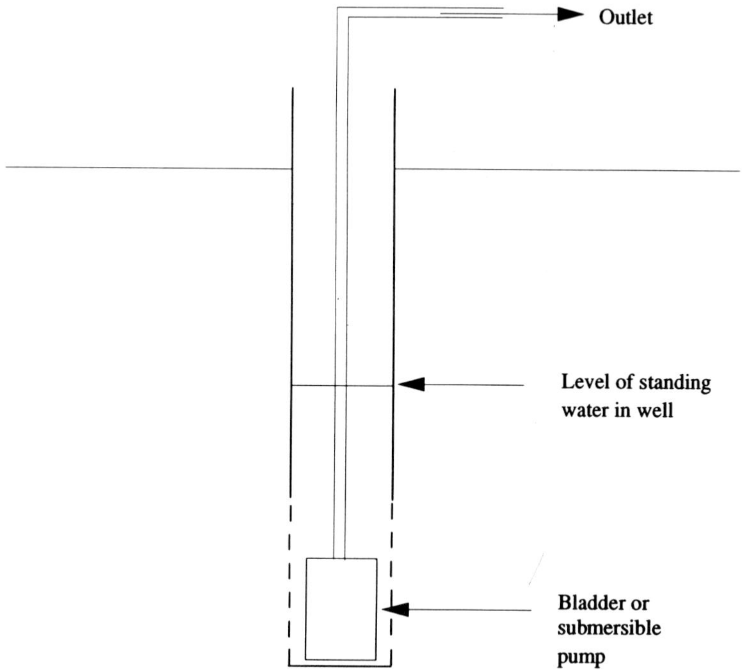

Fig. 2. Sampling a monitoring well using a bladder or submersible pump

Since (as mentioned earlier) most agricultural chemicals for which field research is being conducted to address groundwater concerns are relatively non-volatile, other sampling procedures may be equally acceptable (or even preferred due to the ability to use smaller diameter wells). Pumps located above the ground are often used for collection of samples from wells where the water table is less than about 8 m from the surface. For wells with diameters less than about 4 cm, relatively high capacity (about 1 l/min) peristaltic pumps have been widely used (Fig. 4).

The choice of sampling device may also be influenced by the need to prevent contamination of samples in an environment where mg/kg concentrations may be present in dust and surface soils. One way to help avoid contamination under such conditions is to minimize the introduction of equipment into a well. However, access to the inside of the well must be maintained in order to make water table measurements. For sites where surface pumps are used, one option is to place a relatively small diameter rigid sampling tube permanently inside the well. The cost of foot values used in inertial pumps is low enough to make a similar approach practical. The cost of submersible or bladder pumps may make their dedication to a single well infeasible, especially at sites with a large number of wells. In such situations, equipment placed inside a well must be carefully cleaned before insertion into another well.

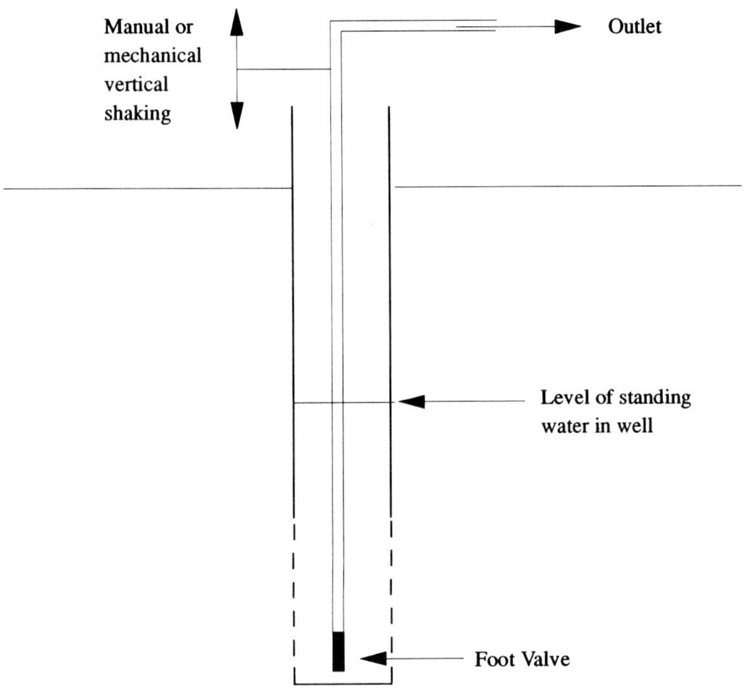

Fig. 3. Sampling a monitoring well using an inertial pump. The power for lifting the water is provided by shaking the tube vertically, either manually or by an appropriate mechanical device

Bailers (Fig. 5) are another device commonly used to remove water from monitoring wells. Advantages of this approach include that any size diameter well can be sampled as long as an appropriately sized bailer is constructed, the equipment is relatively cheap, and no power supply is needed.

Disadvantages are that this procedure can be quite labor intensive when used to purge wells which are screened more than a few meters below the water table and that the continual insertion and removal of any device increases the potential for contamination in an environment where agricultural chemicals may be present in dust or surface soils.

One common use of bailers is in conjunction with other sampling devices. For example, a pump may be used to purge the well and then the actual sample is collected using a bailer to eliminate any concern about materials used in the construction of the pump. However, for a bailer to collect a valid sample, all of the water originally in the well must be removed during the purging process (the studies in Ref. [11] indicate that the water above the intake of the evacuation device may be stagnant and not removed during purging so that either the sample intake must be located at the top of the water standing in the well or the

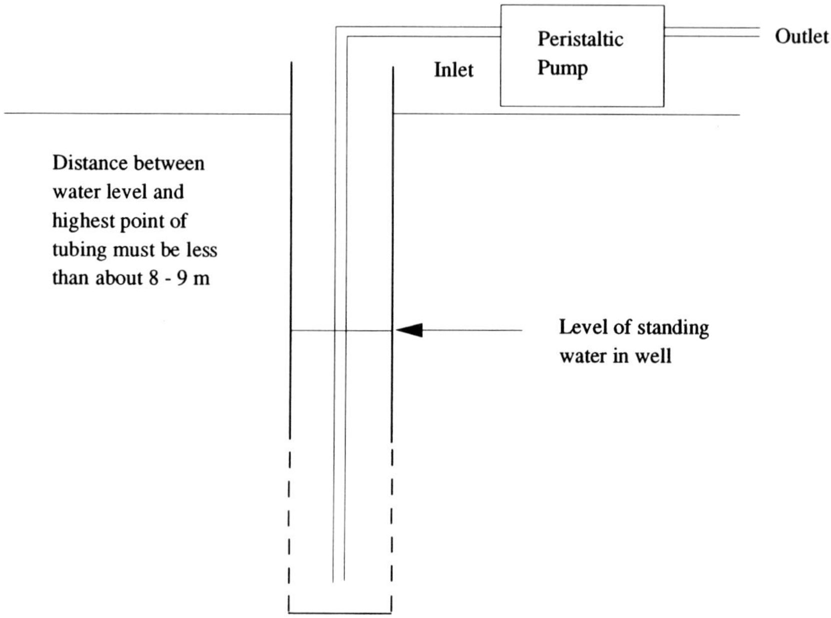

Fig. 4. Sampling a monitoring well using a peristaltic pump

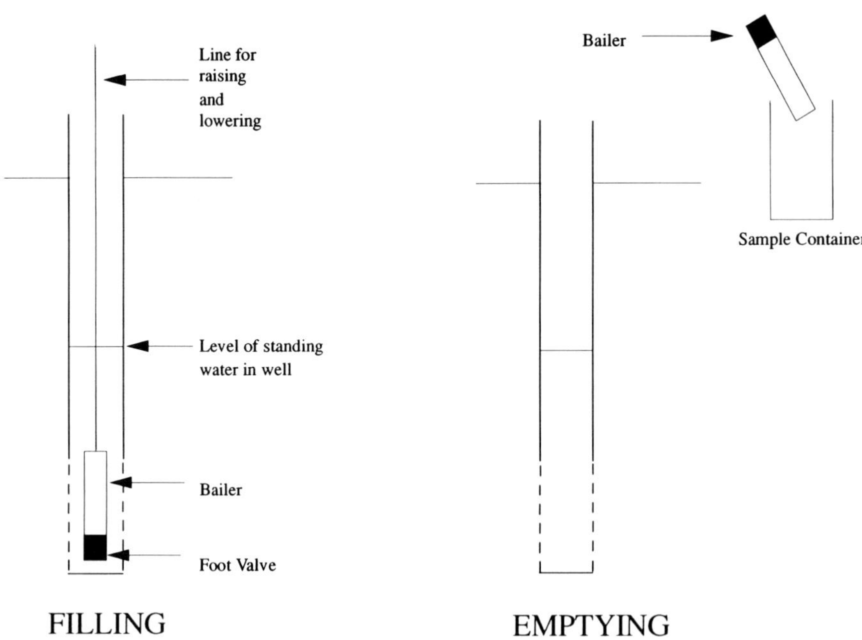

Fig. 5. Sampling a monitoring well using a bailer

pump capacity must be sufficient to completely evacuate all of the water in the well). To minimize problems associated with purging or potentially introducing contaminated dust or dirt into the well, the author's opinion is that whenever possible the sample should be collected at the end of the purging process using the same equipment.

Regardless of the type of equipment or the sampling procedure, careful attention must always be given to cleanliness. Soil must not be introduced into the well during sampling. This may be especially difficult for wells in which the top of the casing is located below ground surface. All equipment should be carefully washed between each well and the wash water should not be discarded in treated fields. Sample containers should be triple rinsed before sample collection. Hands should be kept as clean as possible. All sampling equipment and bottles should be kept off the ground in treated fields and away from dust which might contain residues of agricultural chemicals.

3.2 Potable Wells

Sampling from potable wells is usually simplified by collecting water from an existing tap. Water samples should be collected prior to any water treatment unit, such as a water softener or carbon bed. Since large holding tanks are often present, completely purging the well prior to sample collection is impractical. The procedure used by the author is to allow full flow through the tap for at least a minute, triple rinse the sample container, and then collect the sample.

3.3 Surface Water

In surface water studies, samples are often collected by automatic sampling devices [7]. These devices can be controlled by a variety of sensors to collect samples in a prescribed manner over a runoff event or other time period. Depending on the device, samples can be collected at individual specified time points or a composite sample can be accumulated over a specified time period. Manufacturer's instructions should be followed when using such samplers and care must be taken to use clean bottles and avoid cross-contamination of samples.

In some studies, surface water samples may be collected manually. One simple procedure sometimes appropriate for sampling ditches or ponds is to triple rinse the bottle in the water and then collect the sample by submerging the bottle below the surface. When using this approach, care must be taken not to disturb the water upstream of the sample point (if necessary to enter the water, samplers should approach from downstream of the sample point) and all rinse water should be discarded downstream or away from the sample location.

4 Sample Preservation, Transport, and Storage

After collection, water samples need to be transported to the analytical laboratory without any change in the concentration of the agricultural chemical under study. The most appropriate way of preserving water samples depends on the stability of the agricultural chemical being analyzed. In most studies the common practice of placing samples immediately after collection in an insulated container filled with ice provides adequate stability. However, some compounds can continue to degrade even at 0° C while other compounds are sufficiently stabile in water at ambient temperatures. Occasionally, specific chemical additives may be added to ensure stability (for example, by changing pH or reacting with the chemical). For most agricultural chemicals, data from routine registration studies are sufficient to determine what temperature is needed to provide adequate stability in water, although specific studies may need to be performed with the water from the test site.

If a water sample cannot be shipped immediately to the laboratory, then the sample must be stored at an appropriate temperature until shipment. Usually the samples are kept in a refrigerator, ice, bath, or a freezer, depending on the stability of the compound and the availability of cold storage.

If samples must be shipped to laboratory by commercial freight, then preservation during the shipment period must also be considered. One traditional method is to keep samples cool by packing them in dry ice. Disadvantages of using dry ice include that both plastic and glass bottles packed in dry ice are more easily broken, in some locations dry ice is difficult to obtain, a premium must be paid for shipment, not all commercial carriers accept packages with dry ice, and the chance of a multiple-day delay is increased even with carriers accepting dry ice shipments. One alternative to dry ice is "blue" ice (refreezable blocks sold for use in coolers and lunch boxes).

For most compounds, the author's general practice is to place all samples collected within 20 days of an agricultural chemical application immediately in cooler filled with ice, dry ice, or "blue" ice depending on the relative availability at the sampling site. Samples are then shipped at the end of the day by overnight freight to the laboratory packed with dry or "blue" ice. For many compounds, samples collected at later sampling intervals are placed in a cooler (but not necessarily cooled with ice or dry ice) and shipped at the end of the sampling day without dry ice. Samples not shipped the day of sampling are usually stored in a freezer.

In addition to preventing the degradation of samples between collection and analysis, contamination of samples must also be prevented. Samples should not be transported in the same containers or vehicles used to transport agricultural chemical products. These precautions are especially relevant for studies in which samples are collected immediately before or after an application or for scientists whose jobs include both sample collection and making applications. Similarly, samples should not be stored in containers or freezers also used to store agricultural chemical products. The same precautions are also applicable to sample containers prior to sampling. The author is aware of one occasion in which samples were contaminated by transporting empty sample containers in the trunk of a car which had been previously used to carry products.

5 Record Keeping

Obtaining the necessary sample information and completing sampling records is another important aspect of sampling. The purpose of these records is to identify the sample by establishing the nature of the sample (for example, drinking water, groundwater, or surface water) and when and where is was collected and by whom. Other information may also be needed about the sample location or how the sample was collected. The information required and the forms to be completed varies according to the type and objectives of the study. The following information is based on the general practices of the author.

5.1 Research Programs

Research programs such as field scale groundwater and surface water studies are usually characterized by collection of samples at predetermined sampling points, often at several different dates. Since detailed information on the location of sampling locations is usually already present in study protocols, often the sampling location can be simply identified, for example, D4 (ditch sampling point 4) or C2W3 (well 3 in well cluster 2). Sampling forms (and labels) can be prepared in advance, designating sample numbers and corresponding sample locations. Sampling forms should also include the date (and perhaps the time) of collection and the person collecting the sample. Water table or stream level measurements should be recorded if collected along with any in-field sample measurements such as pH or conductivity. Relevant information on sample collection should also be recorded (for example, the amount of water purged from a well prior to sampling). Figure 6 is an example of a form used for sampling monitoring wells.

5.2 Monitoring Programs

Sampling forms may be quite different for monitoring programs where samples are often easily collected from existing taps which are located by the sampler during the process of sampling. In such programs, gathering needed information about the sampling point may represent more work than collecting the water sample. For example,Table 1 presents information often sought in potable well monitoring programs. A similar approach has been used for monitoring drinking water from surface water.

Rhone-Poulenc Agriculture Field Data

Project Designation: EC-92-025
Trial Designation: GRS-CP1-AT-WW-92
2.0 month post-application groundwater samples
Shipping Form No:
Shipping Date:

Product: compound abc
Formulation: Product 123 EC
Rate: 0.25 kg ai/ha
Crop: winter wheat
Trial Location: Anytown, NL
Cooperator: J. Doe and Sons, Inc.
Treatment: Broadcast
Year of Treatment: 1990

Well Designation	Cluster 3			Cluster 4		
	Well 1	Well 2	Well 3	Well 1	Well 2	Well 3
Sample Number	EB1035	EB1036	EB1037	EB1038	EB1039	EB1040
Sampling Date						
Time of Sampling						
Total Well Depth (m)	3.551	4.786	6.323	3.846	5.037	6.655
Depth to Water (m)						
Standing Water in Well (m)						
Water in Well (L)						
Water to Evacuate (L)						
Water Evacuated (L)						
Sample Temperature (°C)						
Conductivity (μmho)						
pH						
Top of Casing (m bad)	1.138	1.146	1.129	0.824	0.848	0.834
Water Table (m bad)						
Top of Casing to Ground Surface (m)	0.154	0.162	0.145	0.137	0.161	0.147
Water Table (m bgs)						
Well Depth (m bgs)	3.397	4.624	6.178	3.709	4.876	6.508

Abbreviations

bad: below arbitrary datum
bgs: below ground surface
NA: not available
NS: no sample

SOP/Protocol Deviations and Comments: ______

Signature of Sampler(s): ______ Date: ______

Fig. 6. An example of a monitoring well sampling form. Constant information obtained previously is incorporated into the form

Table 1. Typical sample information in potable well monitoring programs

Location Information	Well Information
owner name tenant name well address map and/or other comments	construction (hand dug, drilled, etc.) water table depth well depth casing depth Proximity to storage or mixing/loading facilities
Sample Information	**Application Information**
new or resampled well sample location sample date name of sampler	distance from treated fields application dates and rates crop soil type

5.3 Good Laboratory Practices

When collecting water samples in studies being conducting according to Good Laboratory Practices (GLP), appropriate documentation is essential. All samples must be collected, stored, and shipped according to the study protocol and standard operating procedures, with any deviations appropriately recorded. Sampling and shipping forms and other study records must be completed and signed by the sampler. Chain of custody records must be maintained on all samples from the time of collection until receipt by the laboratory (chain of custody records must also be kept in the laboratory).

Conclusions

Several items must be considered when preparing to collect water samples to be analyzed for agricultural chemicals. These include sample size and container, sample labels, sampling technique, sample preservation from time of collection until receipt by the laboratory, transport and storage of samples, and record keeping. How each of these items is addressed should depend on the characteristics of the agricultural chemical to be analyzed, study objectives and design, and the nature of the water to be sampled. For many agricultural chemicals, requirements for sample containers and sampling techniques can be simplified compared to those often used in other types of groundwater monitoring programs.

Regardless of the specific procedures utilized, sample need to be collected by conscientious and trained personnel. Care must always be taken to avoid contamination by attention to cleanliness of the sampler, sample bottles, sampling equipment, and storage containers. Samples (or empty sample bottles) should not be transported in vehicles also used to transport agricultural chemicals or stored in containers or freezers previously containing agricultural chemicals. Sample labels and study records must be adequate to provide necessary information and fulfil any applicable regulatory requirements.

7 References

1. Nicholson HP, Grzenda AR, Lauer GJ, Cox WS, Teasley JI (1964) Limnol Ocean 9: 310
2. Department of the environment (1991) Drinking water 1990, A report by the chief inspector, Drinking water inspectorate. HMSO, London
3. Holden PW (1986) Pesticides and groundwater quality, Issues and problems in four states. National Academy Press, Washington
4. Leistra M, Boesten JJTI (1989) Agric. Ecosystems Environ. 26: 369

5. Leonard RA (1990) In: Cheng HH (ed) Pesticides in the soil environment: Processes, impacts, and modeling. Soil Science Society of America, Madison, p 303 (SSSA Book Series 2)
6. Jones, RL (1994) In: Honeycutt RC Schabacker D, (eds) Mechanisms of pesticide movement into ground water. Lewis, Ann Arbor, MI p 125
7. Wauchope RD, Decoursey DG (1986) In: Camper ND (ed) Research methods in weed science, 3rd edn. Southern Weed Science Society, p 135
8. Jones RL, Norris FA (1991) In: Nash RG, Leslie, AR (eds) Groundwater residue sampling design. American Chemical Society, Washington, p 165 (ACS Symp. Ser. 465)
9. Kirkland SD, Jones RL, Norris FA (1991) In: Nash RG, Leslie AR (eds) Groundwater residue sampling design. American Chemical Society, Washington, p 214 (ACS Symp. Ser. 465)
10. Simmons ND (1991) In: Brighton crop protection conference weeds-1991, proceedings, vol. 3. p 1259
11. Robin MJL, Gilham RW (1987) Ground Water Monit Rev 7: 85
12. Barker JF, Dickhout R (1988) Ground Water Monit Rev 8:112

Extraction and Cleanup

Malin Åkerblom

Swedish University of Agricultural Sciences, Department of Environmental Assessment, Box 7050, S-75007

Within this chapter numerous techniques and procedures for extraction and cleanup of water samples containing pesticides are presented. These comprise several variations of both the liquid-liquid and the solid phase extraction procedures. Well-known techniques like adsorption chromatography and gel permeation chromatography have been shown to give satisfying results in the cleanup of surface water samples.

Several references are given to critical points in the ultra-trace analysis of polluted water which may contribute decisively to obtaining correct results. Finally, guidelines for newcomers and proposals for extraction and cleaning of water samples are outlined.

Chemistry of Plant Protection 11

1 Introduction

The author's intention in this chapter is to give an overview of extraction techniques as well as to pinpoint steps in the sample preparation and extraction procedures which are crucial for the outcome of the analysis. There are a number of review articles on the analysis of pesticides in water in which sample extraction has been thoroughly discussed, and which give an extended view of the subject [e.g., 1, 2, 3].

The pesticides represent a variety of different chemical characters. They may be very hydrophobic or quite polar neutral organic compounds, they may be weak acids or bases, quaternary ammonium compounds or other ionic substances. Metabolites are in most cases more polar than their parent compounds. Most pesticides have molecular weights between 200 and 500, but there are exceptions. All these compounds cannot be handled by one single method, but many pesticides can be grouped together for part of or for a complete analytical procedure. In a final table, selected references to extraction procedures for about 350 pesticides are given.

The different types of water – drinking, ground, surface, etc. – call for different approaches regarding extraction and cleanup. In surface, leachate, ground and rain water, the pesticides can be present simultaneously in different phases of a water sample [4, 5]. This has been thoroughly discussed by Sigleo and Means [6]. The pesticides can be truly dissolved in the water, they can be adsorbed on suspended particles such as particulate organic matter (POM) or minerals, and they can be bound in different ways to dissolved organic matter (DOM) such as humic and fulvic acids, or other colloids, or as metal complexes. When adsorbed, the pesticides can be transported in the body of water, which has implications for the potential contamination of ground waters [6]. The bioavailability of the adsorbed pesticides, and thus the toxicity, is often but not always reduced [6]. A key question before starting the analysis is therefore whether it is of interest to determine the pesticide content in the three phases, together or separately, or only in some phases.

The two main extraction principles are liquid-liquid extraction (LLE) in which the solutes are partitioned between water and an immiscible solvent, and solid-phase extraction (SPE), in which the water is usually forced or percolated through a column or a membrane in which the solutes are trapped. The extractions are most often performed as a separate step, but for screening, systems on line with HPLC or GC are very useful. The two principles, as well as e.g. alternative evaporation steps, are exemplified by flow charts for the analytical procedures given in Refs. [7] (LLE) and [8] (SPE), see Figs. 1 and 2. Some relevant extraction procedures are summarized in Table 1 and 2. Poole et al [9] produced an excellent review on sample preparation, including extraction and cleanup, both in general and in depth. For overall information on the properties of extraction columns, liquid-liquid extraction devices, and so forth, this review is recommended. It is a general review, not dedicated to pesticides.

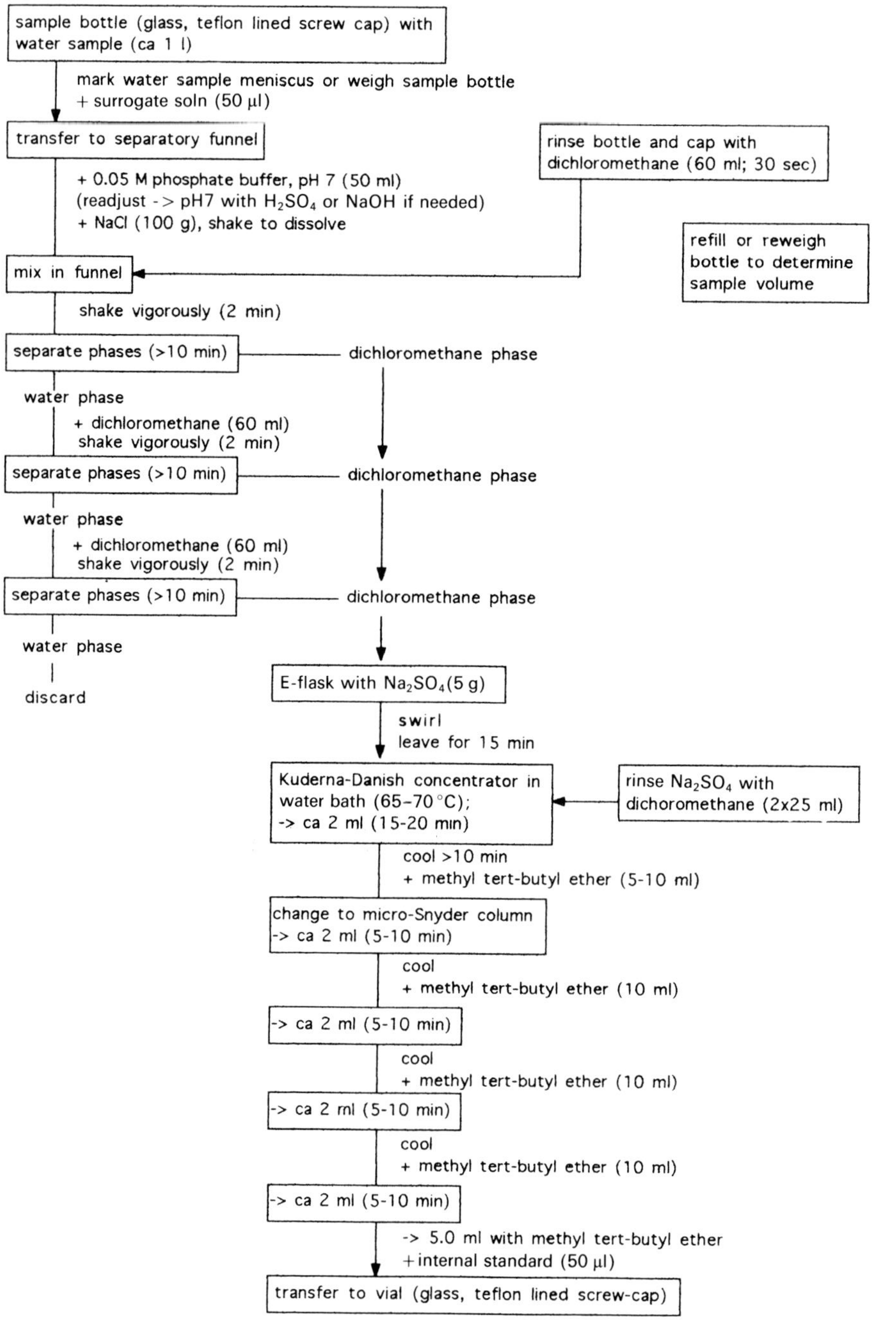

Fig. 1. Flow chart of LLE of semipolar pesticides from, e.g., drinking and surface water (based on Ref. [7])

Table 1. Selected liquid-liquid extraction conditions

Type of water	Volume (liter)	Type of pesticide: number of pesticides	Spiking level ($\mu g\,l^{-1}$)	Solvent (pH adjustment) + NaCl (g)	extraction method: extraction time (min)	Ref no
DW	1	nonpolar: 29	0.01–50	DCM(→7) + 100	tumbling: 60 or funnel: 3 × 2	10
Lab	1	nonpolar: 18	0.01	pentane	magnetic stirring: 10 or rotating perforator: 60	48
SW	0.015	nonpolar: 8	0.5–1.5	hexane	manually: n.s.	60
DW	1	semipolar: 44	0.3–100	DCM(→7) + 100	tumbling: 60 or funnel: 3 × 2	7
Lab, SW	1	semipolar: 11	0.1–2	DCM	rotating perforator: 60	53
GW	0.06	semipolar: 7	0.5–100	hexane + 14	vortex: 1 min	61
DW	1	semipolar: 10	0.025–0.25	pentane : ether (→7.5)	rotating perforator: 120	52
DW, Sea	1	semipolar: 6	0.2–100	DCM : ether acetate : : ammonium formate	n.s.	46
n.s.	1	non & semi: 40	n.s.	pentane	rotating perforator: 30	50
DW	1	non & semi: 25	0.05–2	DCM	funnel: 3 × 20 min	56
SW	10	non & semi: 35	0.015	DCM	continuous (GLSE): 340 ml/min	14
DW	120	non & semi: 6	0.01	DCM	continuous (GLSE): 340 ml/min	57
Well	1	non & semi: 80	1–40	DCM	funnel or tumbling: n.s.	25
Lab	0.25	non & semi: 25	0.1–2	hexane	steam distillation: 90	55
SW	1	non & semi: 85	0.5–10	DCM + 14	funnel: 3 × 2	127
Lab	1	polar & semi: 26	2–50	DCM + 100	funnel: n.s.	132
SW	1	acidic: 14	0.1–1	DCM (< 2)	magnetic stirring: 3 × 30	43
Well	1	acidic: 13	2–10	ether (acidified)*	funnel or tumbling: n.s.	25
DW	1	acidic: 11	0.025–0.25	pentane : ether (→2)	rotating perforator: 120	64
SW	0.2	acidic: 6	1–2	DCM (1–2)*	funnel: 2 × 2	127
n.s.	1	acidic: 8	0.1–10	DCM (< 1)	funnel: n.s.	133

* esters and conjugates first hydrolysed at pH 12
DW: drinking water
SW: surface water
GW: ground water
Lab: laboratory purified water
n.s.: not stated
DCM: dichloromethane

Table 2. Selected solid phase extraction conditions

Type of water	Volume (liter)	Type of pesticide: number of pesticides	Spiking level ($\mu g\,l^{-1}$)	Pretreatment of water	Stationary phase in column (or disk)	Eluent	Ref no
GW	1	non & semi: 9	1–100	+ 0.5% MeOH	C8 disk	ethyl acetate	72
SW, Sea	1–4	non & semi: 11	0.003–0.1		C8	ether : hexane 1 : 1	11
Lab	1	non & semi: 15	0.2–2	→ pH 2 + 0.5% MeOH	C18, or C18 disk	DCM	12
n.s.	0.1	non & semi: 17	0.5		C18	ethyl acetate	85
n.s.	0.1	non & semi: 21	n.s.	→ pH 6	C18	acetone	42
DW	1	non & semi: 58	0.05–0.6	+ 1% MeOH	C8 disk	ethyl acetate	8
Lab, Lab + DOM	1	semipolar: 13	1–200		C18	hexane : acetone 1 : 1	20
Lab	1	semipolar: 45	1		C18 : phenyl 2 : 1	DCM	84
Lab	0.1	semipolar: 11	0.5–20		C18 on line HPLC	acetonitrile : H_2O 3 : 7	22
SW	0.5	semipolar: 7	1		tandem C18/ PRP1 on line HPLC	H_2O : acetonitrile gradient	68
DW	1	semipolar: 11	5		C18	acetonitrile	122
n.s.	1	semipolar: 12	0.05–100	+ 1% MeOH	C18	acetonitrile, MeOH	29
n.s.	0.5	semipolar: 19	n.s.	→ pH 7	C18	acetonitrile	42
SW	0.5	semipolar: 15	0.8		C18	ethyl acetate : isooctane 1 : 1	16
SW	0.005	semipolar: 37	1–5		C18 on line HPLC	H_2O : acetonitrile gradient	98

SW, Sea	1	semipolar: 10	0.1–1		C18	ethyl acetate	27
Lab	0.1	semipolar: 23	25–500		C18	MeOH	132
DW	200	semipolar: 15	0.01–0.1		XAD-2	acetone : hexane 15 : 85	105
DW, SW	0.5–2	1) non & semi: 71 2) acidic: 18	0.1–1.5 0.1–1.5		CpB	1) DCM : MeOH 2) 0.2% TFA in DCM : MeOH 4 : 1	38
DW	0.5	acidic–nonpolar: 19	0.05–1000	+ 22% NaCl	Wofatit Y77	MeOH	94
Lab, SW	0.03	acidic–nonpolar: 38	2–10000	→ pH 3	PLRP-S on line HPLC	H_2O pH 3 : acetonitrile gradient	99
n.s.	0.5	acidic: 2	n.s.	→ pH 2	C18	acetonitrile	42
GW, SW	0.2	acidic: 8	0.1–50		tandem CpB/ SAX	alkaline DCM : MeOH 4 : 1/0.5% TFA + KCl in MeOH	70
SW	0.1	sulfonylureas: 2	0.05–0.2	acidified	C18	H_2O : MeOH 1 : 1	134

DW: drinking water
SW: surface water
GW: ground water
Lab: laboratory purified water
n.s.: not stated
DCM: dichloromethane
MeOH: methanol
TFA: trifluoroacetic acid
CpB: Carbopak B
SAX: strong ion exchanger

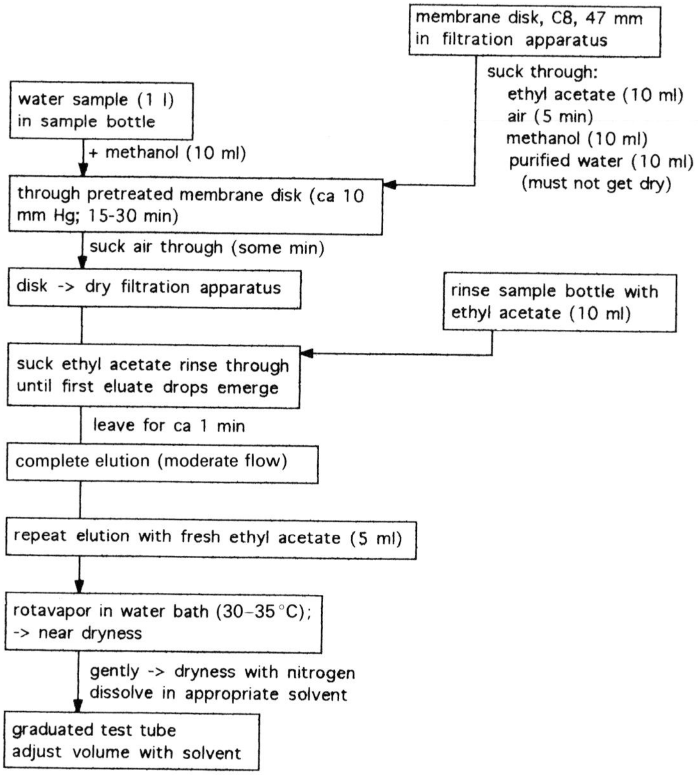

Fig. 2. Flow chart of SPE by membrane disk of non- and semipolar pesticides from drinking water (based on Ref. [8])

In most cases, water samples are brought to the laboratory for analysis. Devices for direct extraction of water samples on site will be described.

2 General Considerations

The extraction of pesticides from water, simple as it may seem with such a "clean" matrix, contains traps not only for the beginner but also for the experienced pesticide residue chemist. The pesticides may be labile, and concentrations very low. Therefore, before the extraction of a water sample, the following points have to be taken into consideration.

2.1 Choice and Pretreatment of Material and Reagents

The suitability of the material and reagents used regarding adsorption, interferences and catalytical properties should always be checked by running blanks and recovery tests at the limit of determination throughout the entire procedure. Purity of all batches of solvents should be checked by concentrating them in the order of magnitude as they are in the procedure. Fractionated redistillation of solvents can provide the purity needed. Solid reagents like sodium chloride and sodium sulphate were freed from interfering organic substances by heating to 400–450 °C for > 4 h [10, 11] or by Soxhlet extraction with dichloromethane [12].

Materials used are in most cases restricted to glass, teflon and stainless steel. These materials are fairly inert, although they may adsorb the most hydrophobic compounds (see below). For cleaning, washing with detergents, thorough water rinsing and final rinsing with an organic solvent is enough in most cases, but this has to be checked. Heating glass wool, glassfibre filters and glass at 400 °C as above may be appropriate, although not for volumetric glass ware. Teflon lined screw caps were extracted over night with methanol before use [10, 7]. When filtration is needed, e.g. for SPE of environmental samples, 0.2–0.7 μm glassfiber filters are most often used.

2.2 Adsorption of Pesticides on Glassware and Other Materials

In some articles, unexpectedly low recoveries of the most hydrophobic pesticides are presented. This may reflect the ease with which the substances may be adsorbed on to the walls or the lid of the sampling/transport/storage bottle or other material (13, 14; Fig. 3). Fenvalerate was thus adsorbed to 54% by the glass bottle in a spiking experiment [11]. Captan was reported to adsorb readily on glass [15]. Polyethylene was found to significantly adsorb organophosphorus pesticides [16] (but not glyphosate [17]). Extraneous transfer of water samples should therefore be avoided. Adsorbed neutral compounds are recovered by rinsing with organic solvents which is then used for extraction [10] or for eluting the solid phase material [8], and ionic species may be recovered with saturated salt solutions.

2.3 Adsorption of Pesticides on Suspended Material and Colloids

As mentioned above, pesticides may be bound to particulate and dissolved matter in natural waters. This causes inhomogeneity in a water sample, which calls for great care [18]. The pesticides most firmly bound to organic matter are the most hydrophobic ones such as the pyrethroids [11, 5, 19] and the organochlorine pesticides [6], but organophosphates and other less hydrophobic pesticides may also be partially bound [20]. Cations, e.g. glyphosate [21], in

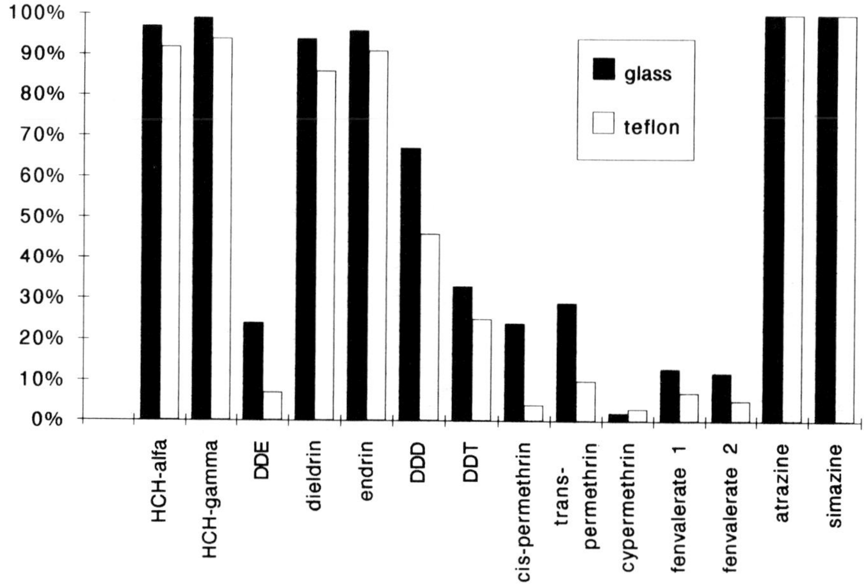

Fig. 3. Adsorption of pesticides to glass and teflon bottles. Fraction remaining in water phase after storage of buffered water fortified with 0.25 $\mu g\,l^{-1}$ of the pesticides indicated. (adopted from Ref. [13])

contrast to anions, are quite strongly adsorbed by ionic forces on minerals, and the quaternary ammonium compounds diquat and paraquat are in addition absorbed within mineral layers [6].

LLE of the entire sample will give the closest picture of what has been transported to the sampled portion of the sampling site. Filtering the sample, as most often necessary in SPE, removes the particle bound fraction. When sampling a river after a heavy rainfall on site with a glassfibre filter – C8 cartridge, Hinckley and Bidleman [11] found 77% of the fenvalerate on the filter. This fraction was evidently adsorbed on particles and thus less bioavailable, which explained the absence of shrimp mortality at the fairly high total concentration found. Marvin et al. [22] extracted benomyl and carbendazim from surface water on a C18 cartridge on line with HPLC. The compounds each gave two peaks when eluted backwards; the greatest peak came from the fraction bound to the particles retained on the protecting 0.5 μm filter, the other from the fraction which had been dissolved in the water and which was retained in the cartridge. They point out that a different filter porosity would have given another result. Freezing and thawing of ground water samples may cause DOM to flocculate and adsorb pesticides, which will then be removed from the sample, if filtered [23].

The fraction of the pesticides which passes a 0.45 μm filter includes pesticides adsorbed to DOM as well as truly dissolved pesticides. In SPE, an appreciable amount of the DOM (mainly humic acids) and the pesticides adsorbed to it may also pass through the stationary phase [4, 20, 24]. Recoveries of organophosphorus and triazine compounds from water fortified with commercial humic acids were reduced by 10–50% [20]. On extraction of a river water with 20 mg l^{-1} DOM, one third of fenvalerate went through a C8 cartridge, presumably associated with DOM [11]. The degree of association with DOM can vary between various natural waters or seasonally. Therefore, when SPE is used for environmental waters, it was advised that the matrix analysed should be spiked with compounds chemically similar to the target ones [20]. The problem seems to be less pronounced with membrane extractions disks, where the lower linear velocity leads to increased extraction efficiency.

Increasing ionic strength would salt out the DOM-compound complex onto the C18 cartridge [6], and lowering pH would retain humic acids on the cartridge [24]. In order to extract all pyrethroid insecticides present, Hadfield et al. [5] packed an extraction column with a layer of strong anion exchanger on top of C8 material. They used this column for in situ extraction of pond water. The DOM together with DOM-bound pyrethroids was adsorbed by the SAX, whereas the free pyrethroids were adsorbed by the C8 material.

2.4 Introduction of a Hydrolytic Step

Many pesticides, e.g. phenoxy acid herbicides, are often sprayed in an ester form and are then hydrolyzed by the plant or elsewhere in the environment to a more active form. The parent compounds, which may also escape to the environment, were converted to the free acids by an alkaline hydrolysis before extraction in, e.g., the chlorinated acids method used by the National Pesticide Survey of drinking-water wells in the USA [25]. Esters of phenoxy acids, ioxynil, flamprop and fluroxypyr were gently cleaved at pH 12 by standing over night at ambient temperature (unpublished results).

2.5 Adjustment of pH

Many authors prefer to standardize pH of the water samples. For example, Schuette et al. [26] assumed that the adjustment of pH (pH 7) and of ion strength (3 M in NaCl) minimized the effects of different matrices on extraction by C18-SPE. Organophosphorus compounds were preferably extracted at pH 6 to 7 due to their hydrolytical instability [16, 27]. Humic acids (major part of DOM), which generally cause considerable interferences, are ionized at neutral pH and hence barely extracted [28], which is normally positive but may be a drawback in SPE. The adsorption to DOM may vary with pH. Thus the K_{oc} of atrazine increased four-fold from pH 5 to pH 8 [6].

In LLE, the pH needed for efficient (or selective) extraction of weak acids and bases is dependent both on the pK_a of the substance and of the extraction

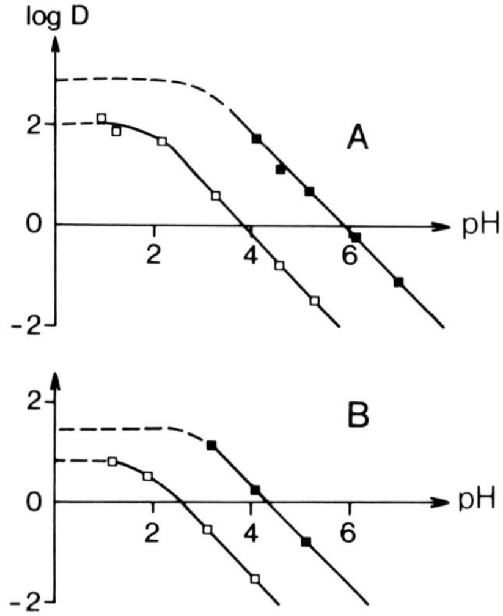

Fig. 4. Distribution (D) of 2,4-D (■; pK_a 2.8) and picloram (□; pK_a 2.0) between ethyl acetate (**A**) or dichloromethane (**B**) and phosphate buffer as a function of pH. Equal phase volumes, single extractions. [135]

power of the solvent used, which is illustrated in Fig. 4. The same principle is partly true for hydrophobic SPE as well, although ionized molecules may well be retained by a C18 sorbent provided the apolar part is large enough [29, 30].

2.6 Adjustment of Ionic Strength

The recovery of the more polar compounds both in LLE and SPE may be increased by increasing the ionic strength, which weakens the solute-water interaction [31, 7, 28, 32]. This effect is more pronounced, the more polar a compound is [33]. A strong salting out effect was found in the C18 solid phase extraction of polar organophosphorus pesticides, whereas a decrease in recovery was discussed for the most hydrophobic ones [16]. Dedek et al. [34], applying SPE on Wofatit Y77, enhanced the recovery of the very polar compound sodium trichloroacetate from 22 to 89% by saturating the water sample to 80% with sodium chloride. The possibility of salting out water miscible solvents like acetonitrile and 2-propanol and thus of extracting very polar solutes was demonstrated by Leggett et al. [35].

2.7 Losses by Evaporation or Degradation

Usually the extract obtained has to be concentrated before the final determination. Larger volumes are evaporated at a slightly elevated temperature in rotavapor or Kuderna-Danish devices down to some ml, cf. Figs. 1 and 2. The final evaporation is preferably performed in a micro-Kuderna Danish device [3] or in a test tube or vial by a very gentle stream of pure nitrogen (preferable to air which may oxidize compounds). All residues on vessel walls or occluded in coextractives must be thoroughly dissolved, preferably by ultrasonic treatment.

Evaporation is a risky procedure. The loss of pesticides on evaporation was systematically investigated [36], and it is undoubtedly reflected in some reports on low recoveries of pesticides eluting early in gas chromatography. The addition of a minor amount of a keeper such a dodecane [36], polyethyleneglycol or OV101 [37] prevents losses appreciably and is vital if the extract has to be completely evaporated. Transesterification of fluazifop-butyl in the presence of methanol [29], and methylation of coumafuryl and warfarin on evaporating an acidic methanolic extract [38] may occur when the extract becomes concentrated and heated. If a change of solvent is necessary, it is preferably performed gradually [7, 10, 39].

Pesticides may have been oxidized by chlorine or hypochlorite in treated drinking water [7, 12] or by unknown oxidative species in ground and river water (38). This is prevented by addition of sodium thiosulfate [7], or preferably sodium sulfite [12, 38] since sodium thiosulfate may produce a residue of elemental sulfur, which may interfere with some analytes. On storage of extracts, susceptible pesticides may be hydrolysed by minor water residues unless the extract is dried by an agent such as sodium sulphate.

2.8 Quality Assurance

Norris gave crushing evidence of the need for quality assurance programs in his test of ten randomly selected contract laboratories, getting both false positive and false negative results [40]. Similarly, 6 out of 11 laboratories reported atrazine in atrazine-free water [41]. An in-depth discussion on fundamental QA principles is found in the last Chapter.

When setting up a method, recovery tests at relevant levels should be performed with all substances and water types of interest. Since it is then usually not feasible to carry out regular recovery tests on a routine basis with all compounds targeted in a multiresidue study, a battery of pesticides should be selected. This test battery should include compounds representing the most and least polar ones, the most volatile compounds and the substances most susceptible to oxidation and other types of degradation. All materials and reagents as well as ambient air and compressed gases should be checked for impurities by regular analyses of clean reagent water. A field reagent blank consisting of reagent water in a sample bottle, which is treated in the field as true samples in all respects, is included in the EPA monitoring methods [12]. Sources of interference and bad recoveries are most easily identified by working stepwise backwards through the procedure. Strictly kept records help to trace contamination and memory effects which may arise from heavily polluted samples.

The EPA/AOAC methods for organochlorine [10] and nitrogen- and phosphorous containing pesticides [7] stipulate the addition of both 'surrogate' and 'internal' standards to each sample. The surrogate is added to the sample prior to extraction, and is used to check the overall recovery. The internal standard is added to the final sample extract to compensate for fluctuating chromatographic performance including injection volumes and detector response. The internal standard is also used to compensate for inaccuracy when handling low volumes of final extracts [18, 42].

Several authors emphasize that, due to the unpredictable nature of matrix effects in natural waters, the matrix effects should always be determined [16, 20], preferably with a battery of surrogate standards [14]. Alas, most recovery data in the literature is obtained from extraction of reagent or tap water, even in those cases where the method is routinely used for environmental samples, a deficiency which was stressed by Johnson et al. [20]. Quality control charts should be set up both for surrogate standards [12, 43] and for the regular recovery tests [44]. It is advisable to have different control charts for different types of water.

When fortifying a sample with surrogate standards or for recovery tests, the solutes should be dissolved in a solvent which is miscible with water in the quantities used, in order to be immediately dispersed in the water. The risk of subsequent adsorption of hydrophobic pesticides on the glass or teflon vessel has been discussed above.

The effect of storage of samples with preservatives at + 4 °C and of final extracts (temperature not given) was extensively tested for 147 compounds, cf. Ref. [25]. After 14 days substantial losses were noticed for 26 compounds in the water samples and for 4 compounds in the extracts [45].

3 Liquid-Liquid Extraction (LLE)

Among the solvents used for the extraction of pesticides, ethyl acetate and dichloromethane are among those preferred, due to their ability to dissolve and extract compounds with a wide range of polarities. Dichloromethane dissolves less water, is more easily evaporated and is nonflammable, whereas ethyl acetate is a better solvent for more polar solutes, see for example Ref. [46], and is less toxic and more environmentally acceptable. The extraction efficiency is modified by adjustment of pH and ion strength, as discussed above. Alkanes such as cyclohexane or pentane are suitable for selective extraction of the most hydrophobic pesticides [47]. One advantage of these solvents is that they are neither partly dissolved in nor dissolve water. Their ability to extract the pesticides from material present in natural waters must be checked, though. Very polar solutes may be extracted with salted out polar solvents, see above [35].

3.1 Off-Line LLE, Medium Scale

The classical way to perform liquid-liquid extraction is by separation funnel extraction. A vigorous funnel shaking creates a large contact area between solvent and sample, which leads to a high rate and degree of extraction. Mechanical shaking reduces the inevitable variability obtained when shaking by hand as well as occupational health risks. Common alternatives to funnel shaking are prolonged tumbling of a flask end over end [7, 10], and magnetic stirring [43, 48, 47]; cf Table 1. Vigorous mixing of solvent and water, especially surface and leachate water, may create emulsion problems. With repeated extractions, the emulsion should stay with the water phase until the last extraction [10]. The emulsion may then be broken by filtering through a plug of glass wool [47], by centrifugation, refrigeration, salting out or by addition of a small volume of organic solvent [9].

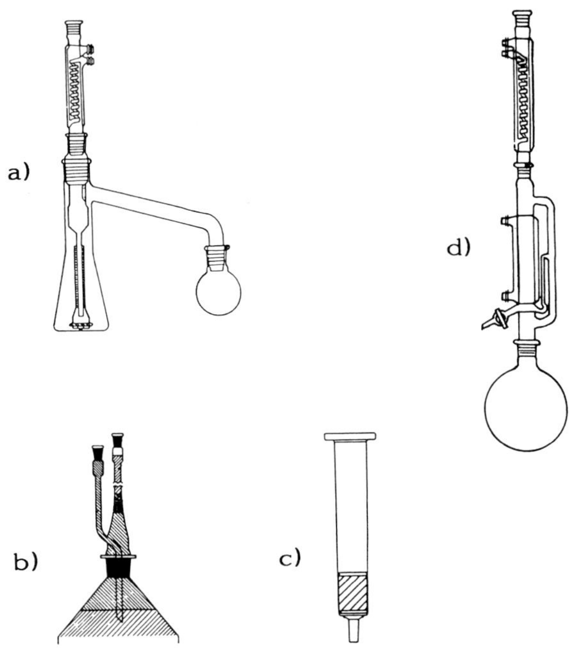

Fig. 5a–d. Extraction devices. **a** Rotation perforator. Solvent in round bottom flask is heated to evaporate, is condensed and then distributed into the water sample in conical flask as small droplets by the rotating distributor, extracting the pesticides on its way back to the solvent vessel. **b** Microseparator attached to a bottle after, e.g. magnetic stirring extraction. Purified water is added through the side arm to force the solvent into the narrow part of the connected tube. **c** Solid phase extraction cartridge. **d** Extractive steam distillation apparatus. The water is refluxed, and the condensed steam percolates cooled solvent, where the codistilled pesticides are extracted. ([48]; reproduced with permission from Pergamon Press)

A continuous liquid-liquid extractor was used for the extraction of organochlorine pesticides [49]. In this cyclic system, the solvent is percolated through the water sample. The solvent of the extract so obtained is evaporated, condensed and passed through the water again. In a similar way, the technique of rotation perforation extraction was applied to non- and semipolar and acidic compounds with pentane [50, 48] or pentane:diethyl ether 1:1 [51, 52] as solvent; cf Fig. 5a. By using dichloromethane, van Hoof et al. [53] succeeded in extracting the more polar herbicides chloridazon and metoxuron, which previous authors had failed to extract. As a rule low boiling solvents should be chosen for continuous LLE in order to avoid elevated temperatures of the extract.

Several devices are available for simultaneous steam distillation and solvent extraction ([9]; cf Fig. 5d). The solutes might also be trapped in a suitable sorbent column instead of a collecting solvent [54]. Within the pesticide field, this has mainly been applied to heat stable compounds as organochlorines and fumigants and to phenolic compounds [55]. The technique was investigated for nearly 40 pesticides, with acceptable recoveries mainly for organochlorines. Legrand et al. [56] applied alkaline hydrolysis of urea herbicides and simultaneous steam distillation extraction of their corresponding substituted anilines. Recoveries were 50–80% at the levels of 0.05–0.2 $\mu g\,l^{-1}$. When analyzing samples of unknown origin it must, however, be taken into account that the anilines, like chlorophenols, can stem from sources other than pesticides.

3.2 Large-Scale Extractions

In some environmental investigations, especially of rain water or of persistent organochlorine pesticides, very low concentrations are targeted. One way to achieve this is to extract large volumes of water. Nine model pesticides were extracted with the so-called Goulden Large Sample Extractor, GLSE, in which 120 l of water were continuously pumped into the bottom of an extraction vessel and were dispersed in the 200 ml dichloromethane present in the vessel by means of a stirrer [57]. Substances with a partition coefficient > 1000 between dichloromethane and water were satisfactorily recovered. This device was validated for the screening of 35 pesticides from stream waters [14], for ng l^{-1} levels of organochlorine pesticides in snow, in which case 40 l of melted snow was pumped through on site [58], and on board a monitoring ship on the Great Lakes, Canada/USA [2]. The Aqueous Phase Liquid Extractor (APLE) is capable of in-field extraction of 200 l of water by recirculating 8 l of high density solvent via a spray-bar over a prolonged period [59].

3.3 Small-Scale Extractions

There is a waste of sample and/or loss of sensitivity in conventional residue analysis where only a small fraction of the final extract, for GC some µl, is taken for determination. By using a concurrent solvent evaporation technique, van der Hoff et al. [60] were able to inject 200 μl extract into a GC. In this way they reached ppt-ppb levels with only 15–25 ml of water sample. No extraction was needed when 400 μl of the well water samples were injected onto HPLC in the National Pesticide Survey study. A sensitive post column derivatization made the determination of carbamates down to the level of 1–2 $\mu g\,l^{-1}$ possible [25]. Potter et al. [61] sampled column leachate or ground water in serum bottles, fortified with sodium chloride for preservation and salting out effect. One ml hexane was added through the septum using a syringe and the sample was vortexed. Absolute recoveries of the acetanilide herbicides and their degradation products were in the range 45–74% but reproducible for each compound, which allowed for correction against an internal standard.

3.4 On-Line LLE

The flow injection extraction technique was investigated for automated LLE-GC screening of HCHs in ground water [62]. The water was segmented by extracting solvent and passed through a 6 m extraction coil. After passing a phase separator, the extract was further passed on line to a GC for concurrent solvent evaporation injection. Similarly, Farran et al. [63] successfully investigated the possibility of segmented flow extraction coupled with HPLC for some phenoxy acid herbicides and organophosphorus insecticides in the 1 $mg\,l^{-1}$ range. By altering the detection mode (MS full scan to MS SIM) they predict the application in the $\mu g\,l^{-1}$ level.

4 Solid Phase Extraction (SPE)

In SPE, the water, with or without modifier, is usually forced through a preconditioned column or filter with a regulated flow, either by vaccum or by pressurized nitrogen or air, or attached to a pumping system, off line or on line with a chromatographic system. A thorough presentation of the solid phase column extraction technique, with its pros and cons, was given by Liska et al. [19], including a description of sorption and desorption processes and of various sorbent materials. Van Zoonen et al. [64] discussed the use of coupled chromatographic techniques in pesticide residue analysis.

Among the number of sorbents that have been used for the extraction of pesticides from water, the modified silicas, the macroreticular resins and graphitized carbon black will be discussed below. The modified silicas are at present the most commonly used materials, especially those with chemically bonded C_{18}- and C_8-carbon chains, here referred to as C18 and C8. These are best suited for non-polar and semipolar compounds. Macroreticular resins, notably XAD-

2, XAD-4 and XAD-7 have long been used, covering the same range of polarities as the modified silicas. Graphitized carbon black, represented by Carbopak B, has been applied to a broad range of pesticides including both polar and nonpolar ones, and has greater affinity for acidic compounds than the modified silicas. The polymers PRP-1, PLRP-S and Wofatit Y77 are especially suited for more polar compounds. Ionic species can be extracted with ion exchange material when strongly retained as glyphosate [65] and the diquaternary ammonium compounds diquat and paraquat [66]. Less strongly retained ions will have to compete for the binding sites with mineral salts present in the water samples [67, 68]. Such substances, however, can first be extracted from the water by a hydrophobic column and then selectively adsorbed on an ion exchange column, attached in tandem either off line [69, 70] or on line with a chromatographic system [68].

4.1 Off-Line SPE

4.1.1 Modified Silicas

In numerous methods the modified silicas have been employed, cf Table 2. C18 sorbent is used in the official EPA Method 525 [12]. Within ISO, a strategy for the use of extraction columns for semipolar compounds is being prepared [71]. The material is available prepacked in, e.g., polypropylene cartridges from several suppliers or in bulk for home made columns. Most studies reported have used cartridges packed with 0.1–3 g of 30–60 μm particles. Membrane extraction disks are a promising alternative to SPE cartridges. They have been used for nonpolar to polar pesticides [8, 37, 72–74] as well as for acidic compounds [75] and were validated for the compounds included in the EPA methods 507, 508 and 525.1, cf [7, 10, 12, 76]. The disks consist of fibrillated PTFE into which about 0.5 g of 8 μm adsorptive particles are enmeshed.

The performance of SPE cartridges varies very much, both between those from different suppliers ([28]; Fig. 6) and between batches. So does the abundance of monomers, plasticizers and other impurities from the adsorbent, cartridges, frits and packaging plastic bags ([26, 77]; Fig. 7). EPA [12] warns that several sources of SPE cartridges should be evaluated before an acceptable supply is identified. A prewash with the elution solvent, or Soxhlet extraction [11], removes contaminants from manufacturing or environmental exposure. Home packing in glass columns circumvented most of the problem [78], but the risk for channelling might be greater.

Environmental waters have to be filtered to prevent clogging of the adsorbent. The particle bound pesticides will then be lost unless the residues on the filter are extracted, as discussed previously. Filtered samples containing iron hydroxide or considerable amounts of microorganisms, still clogged the SPE cartridges [48], and inorganic salt particles plugged the pores of membrane disks [72]. These particles dissolved when the water was acidified. A Durapore

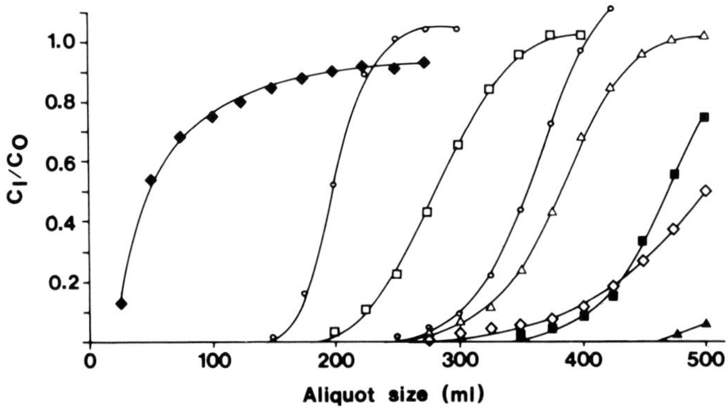

Fig. 6. Breakthrough curves of fenuron on 1 g C18 disposable cartridges from different producers, showing the fraction of fenuron which has passed through after a given volume of water extracted. ([28]; reproduced with permission from Hewlett-Packard)

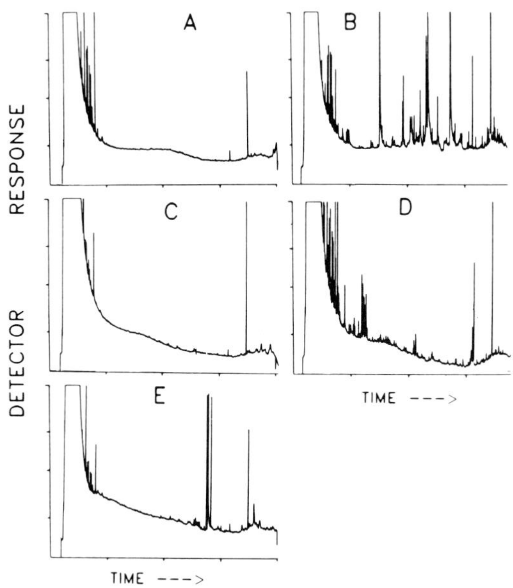

Fig. 7. Impurities revealed in C18 material of disposable cartridges from different producers. The material was extracted with benzene and the extract analysed by GC-FID. ([77]; reproduced with permission from Anal. Chem.)

hydrophilic 0.45 μm filter proved effective in preventing the disk from being plugged with clay particles [37]. The risk of losing pesticides associated with dissolved organic matter (DOM) has been discussed above.

The simplest way of transferring the sample from the sample bottle to a cartridge is to immerse the thoroughly cleaned cartridge in the water sample

[79, 80] or to attach it to a pipette [81] or teflon tubing, which is put in the water sample, and draw the sample through. More commonly the cartridge is attached to the outlet of a separating funnel containing the sample [12, 82]. Membrane disks are placed in a Millipore filtration apparatus and vacuum is applied to give the desired flow [12, 72].

Prewetting the stationary phase with solvents such as methanol, isopropanol, acetonitrile or acetone opens the hydrophobic chains, and thus increases the accessible surface area [19, 29]. As large amounts of water sample are passed through the column or disk, the prewetting solvent is eventually removed, which leads to inconsistent results [83], especially for hydrophobic solutes. The water sample is therefore often fortified with about 1% of methanol or isopropanol so that the stationary phase is continuously wetted. However, too polar compounds, like the dealkylated triazine metabolites, may be retained less in the presence of modifier [28, 71].

The flow rate, or more precisely the linear velocity, of the water being extracted may influence the equilibrium of the solvent-solute system [19]. An increase in extraction efficiency with lower velocity was observed especially for more polar, early eluting compounds like propoxur and carbofuran [15] and for pesticides associated with DOM [24]. The EPA Method 525 for C18 cartridge extraction [12] gives a 2 h extraction of 1 l of water, which gives a linear velocity of 0.18 $cm\,s^{-1}$ with the cartridge used (as calculated by Hagen et al. [72]). The modified EPA Method 525, using membrane disks, recommends a flow of 1 l in 20 min [12], which due to the smaller particle size of the membrane gives a linear velocity of 0.07 $cm\,s^{-1}$.

Many authors prefer to remove residual water prior to elution in order to get the eluting solvent in close contact with the stationary phase, as well as to produce a water free extract for the subsequent determinations. This is achieved by pressure or suction of dry nitrogen or air through the solid phase for some minutes to several hours, by freeze drying [28], by storing in a desiccator or by centrifugation [19]. If the water is not eliminated by drying the column, the eluate may be passed, in series, through a small column of sodium sulfate [26], or the water is separated by freezing the dichloromethane eluate [84] or evaporating ether from the ether : hexane eluent [11].

For the elution of the pesticides a variety of eluents are in use, as can be seen in Table 2. The final determination step partly governs the choice of eluent. Methanol and ethyl acetate seem to be the most universal. Ethyl acetate, in contrast to hexane, fully eluted triazine herbicides and organophosphorus insecticides from C18 columns [27, 78]. Dichloromethane was chosen in the EPA Method 525 for elution of non- and semipolar compounds from C18 cartridges and membrane disks [12]. More than 90% of almost all pesticides were predicted to be eluted with ethyl acetate in the first 60 μl eluate from a 100 mg C18 column, where ml amounts of methanol or acetonitrile are needed [85]. For the elution of more hydrophobic organophosphorus insecticides ($\log K_{ow} > 4.9$) from the C18 column used, neither methanol nor ethyl acetate was sufficient but a mixture of ethyl acetate and isooctane (1 : 1) was [16].

Acetonitrile, acetone and ethyl acetate were more selective than methanol towards impurities from river water, and propham, chlorpropham and crimidine were claimed to be partially decomposed by methanol [28]. Recently, SFE (supercritical fluid extraction) was tested for the extraction of sulfonylurea herbicides from membrane disks [86] and hydrophobic compounds including organochlorine pesticides from C18 cartridges [87]. Until now, there seems to be no obvious advantage of SFE over conventional eluents for modified silicas.

4.1.2 Graphitized Carbon Black

The suitability of Carbopak B was thoroughly investigated by DiCorcia, Marchetti and coworkers [38, 69, 70, 67]. In their method for drinking, ground and river water, DiCorcia and Marchetti [38] successfully included 89 pesticides, 18 of which were acidic compounds. The Carbopak B used was coarse enough to allow even river samples to be extracted unfiltered. Excellent recoveries were reported when 2 l of tap water spiked at the 0.1–1.5 $\mu g\,l^{-1}$ level were extracted. Due to matrix effects, the volumes were reduced to 0.5 l for river water and 1.5 l for ground water. In earlier work, triazines [69] and phenoxy acid herbicides [70] were selectively extracted by putting ion exchange sorbent in series after Carbopak B. The solutes were first trapped in the Carbopak B. On elution, they were selectively adsorbed on the ion exchange sorbent. Mangani et al. [82] obtained good recoveries of organochlorine insecticides and some herbicides from 1 l of water with 50 mg of Carbopak B at the level of 5 $ng\,l^{-1}$.

4.1.3 Macroreticular Resins

The XAD resins have long been used for the extraction of a variety of pesticides. XAD-1, XAD-2 and XAD-4 are styrene-divinylbenzene copolymers with a hydrophobic character. XAD-7 and XAD-8 are acrylic ester resins with higher affinity for polar solutes. XAD resins are known to release inteferences and need pretreatment by extensive Soxhlet extraction [88, 89], or through preconditioning of the packed column [90]. The material has often been used for in situ extractions of large volumes. Mattern et al. [90] combined XAD-2 and XAD-7 in the same cartridge for the extraction of semipolar and quite polar pesticides from river water.

The styrene-divinylbenzene-copolymerisates PRP-1 [68, 91, 92, 93] and Wofatit Y77 (= Wofatit EP63) [34, 94, 95] are especially suitable for the more polar pesticides. E.g., methamidophos and trichloroacetate ion (water solubilities > 2 and 1.2 $kg\,l^{-1}$ respectively) were recovered by Wofatit Y77 to about 90% from 0.5 l of drinking water, 80% saturated with sodium chloride [34]. The more hydrophobic pesticides are firmly bound and were preferably adsorbed onto a C18 precolumn [93]. Otherwise a temperature for the eluting methanol of 60 °C was recommended [95].

4.2 On-Line Procedures

Rapid screening of samples is achieved by coupling an enrichment column in series with a subsequent chromatographic determination. The water sample is pumped through a precolumn, in which the solutes are trapped, and the water goes to waste. The precolumn may then be selectively cleaned with a small volume of a suitable solvent mixture. By a system of switching valves, the solutes are then eluted onto the analytical HPLC column or into a GC retention gap. When feasible, the on-line procedures have several advantages. They require few manual steps and they are very well suited for automated screening. In off-line SPE or LLE procedures normally only a small fraction of the final extract will be used. In on-line systems, the whole amount of solute extracted is used for injection into the chromatographic system, which allows reduced sample volumes. However, in the case of positive finds, a new extract has to be made to verify the find by other means. Much of what has been discussed previously concerning, e.g., choice of sorbent, risk of losses on filtration and by DOM association, pH and ionic strength adjustments, is valid for on-line procedures as well.

In the simplest set up the loop of the manual injector of a HPLC was replaced by a short C18 precolumn and 2–16 ml of the acidified and sodium chloride saturated sample of phenoxy acid herbicides was applied manually with a syringe [33]. A PRP-1 stationary phase combined with an extra pump for sample delivery was found to be more efficient. Geerdink et al. [91] thus extracted phenoxy acid herbicides on a 10 mm PRP-1 precolumn from up to 100 ml of acidified drinking water. After extraction, an efficient clean up was achieved by slowly washing the precolumn with only 4 bed volumes (140 μl) of acidified acetonitrile-water (3 : 7) prior to desorption with the HPLC eluent.

A fully automated method was set up for the screening of triazines and their desethyl metabolites, and terbutryn and metolachlor in ground and tap water [96]. In each series, 32 samples were extracted on a C18 precolumn and analysed by HPLC.

Benomyl is rapidly converted to carbendazim by organic solvents. In order to distinguish between these two fungicides in water, the quick procedure of C18 on line extraction was successfully applied to 100 ml samples [22, 97]. Difenzoquat, a quaternary ammonium ion herbicide, was extracted from up to 48 ml of water on a 7 cm C18 precolumn [30].

A beautiful application of extraction on line with HPLC was demonstrated by Reupert et al. [98] in their screening method within the Rhine Basin Program, Fig. 8. About 40 polar and semipolar pesticides were selected by production volume and history in water quality monitoring for inclusion in this method. A level of 1 $\mu g\,l^{-1}$ is targeted, which is reached with a 5 ml sample. In the search for suitable column enrichment material, the least retained compound, desisopropylatrazine (= desethylsimazine) was extracted by 14 different C18 and C8 columns of equal size, giving breakthrough volumes between 1.4 and 11 ml. As they point out, in on-line techniques the peak broadening is

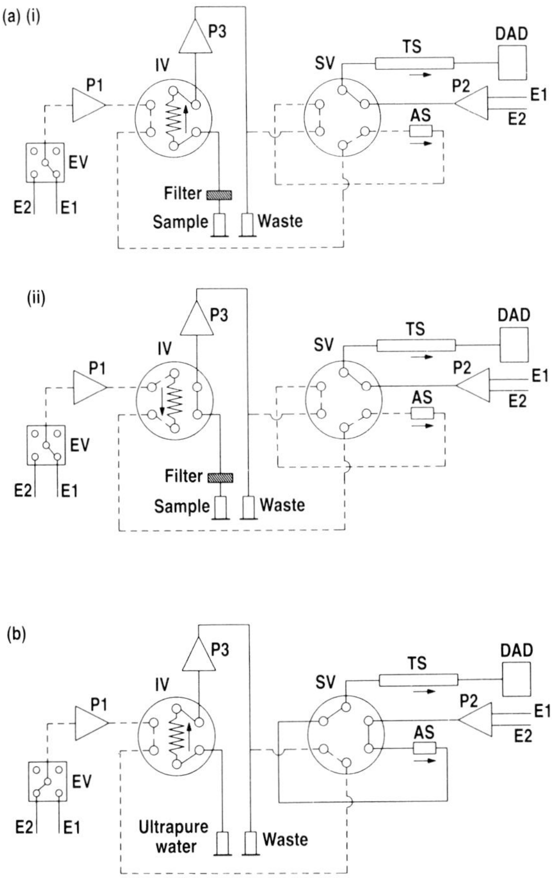

Fig. 8a, b. Automated solid phase extraction on line with HPLC. **a** i: sample intake: ii: SPE **b** transfer to analytical column and analysis. IV = injection valve with 5 ml loop; P1, P2, P3 = pumps; E1, E2 = eluents, EV = solvent switching valve; AS = extraction column; SV = column switching valve; TS = analytical column; DAD = diode array detector ([98]; reproduced with permission from LC-GC Intl.)

a limiting factor, and 3–10 μm particles are chosen. A similar procedure was applied by Liska et al. [99] for the screening of over 50 pesticides in river water by preconcentration of 30 ml of water on a PLRP-S column.

Small membrane extraction disks were placed in a homemade membrane holder, which was put in line with an HPLC [73, 74] or GC [100] system. The reduced amount of C18 material was compensated for by stacking membranes in the holder. The system was applied to the extraction of polar and semipolar pesticides from tap and surface water.

C18 precolumns were put in tandem with PRP-1 precolumns for the simultaneous extraction of non polar and relatively polar solutes. The two precolumns were then separately eluted onto the analytical column by means of switching valves [68, 92, 93]. When 50 ml of water were extracted, all of the 6 urea herbicides tested apart from metoxuron and buturon were fully retained by the C18 column. Increasing sample volume to 500 ml caused all herbicides to pass to, and be retained by, the PRP-1 column to a great extent [92]. The ratio of amounts preconcentrated on the two columns was judged to be useful information for identification [92]. The possibility of using the repeatability of the automated system to allow extraction of larger sample volumes than breakthrough volumes was discussed [93] and was later applied in a similar system [99].

In a very interesting approach, an optical fiber coated with some GC stationary phase was glued to a stainless steel rod. This rod ran up through a Hamilton syringe. The fibre was dipped into the water sample for a specified period of 2–15 min, then drawn up into the syringe needle, and was thermally desorbed in the injection port of a gas chromatograph [101, 102]. The device has recently become commercially available. Other approaches include the application of passive dosimetry in water [103], and the extraction of a water sample with the stationary phase in a capillary column through which the sample is forced, and which is subsequently put into a gas chromatograph [104].

5 On-Site Extraction

The transport of water samples is often troublesome. It also limits sample volumes, and involves the risk of degradation of the pesticides and adsorption of hydrophobic pesticides onto the sample vessels. On-site extraction circumvents these limitations and is, in addition, suitable for continuous sampling of, e.g., a stream or rain water.

Solid phase extraction was applied to 100 ml samples by forcing the sample through a 100 mg C18 column attached to a syringe [85]. Hinckley and Bidleman [11] sampled 1–4 l of sea and river water on site by pulling the sample through a glassfiber filter-C8 cartridge train with a perstaltic pump. However, they lost DOM associated pyrethroids, which was avoided by Hadfield et al. [5], who sucked the water through a SAX/C8 column with a hand-held vacuum pump for pyrethroid determination.

Lebel et al. [105] reached levels of 1 ng l^{-1} when monitoring drinking water for organophosphorus pesticides by attaching a XAD-2 column to the water tap for 24–48 h. Green et al. [106] placed a XAD-2 column in an in situ sampler for extracting organochlorines from sea water at a depth of 20 m. Within 3 days, about 100 l of water was pumped through. The previously mentioned large scale LLE extractor, GLSE [14], was used for the extraction of 50 l samples during monitoring cruises in a shipboard laboratory [2]. Highly hydrophobic compounds were extracted on site in a continuous LLE device from hundreds of liters of river or sea water into 100–300 ml of apolar solvent [107]. It was found preferable to XAD-2 extraction for some compounds in the presence of DOM [108].

Alberti and Willemsen [88] continuously extracted non- and semipolar pesticides from river water by sucking water through a 15 cm column of XAD-4, with a particle size of 0.3–1 mm. The column was replaced every 12 h. Strachan and Huneault [109] used cartridges with XAD-2 for in situ extraction of pesticides from rain water.

An elegant method for in situ enrichment of phenoxy acid and sulfonyl urea herbicides was presented by Nilvé et al. [110–112] and was applied by Mathiasson et al. [113]. They passed a continuously acidified water stream over a Teflon membrane which had been impregnated with *n*-undecane and di-*n*-hexylether, and which separated the water stream from a neutral buffer. The non-ionized herbicides dissolved in the membrane phase and passed through the membrane to the buffer, where they dissociated and were no longer prone to dissolve in the membrane phase. In the laboratory, the cartridge was attached on line to the HPLC and the herbicides were eluted onto the analytical column.

6 Methods for Single Pesticides

Paraquat and diquat may be extracted at low concentrations by a cation exchanger batchwise [114] or on a column [66]. If the sample contains suspended material it will remain on the filter on top of the column. The more bioavailable fraction which is loosely bound by ionic forces to the particles, will be desorbed with the eluting strong salt solution. The fraction which is firmly bound within the mineral layers has to be freed by hot acid disintegration of the solids [66]. Silica gel column extraction was applied for dissolved diquat, paraquat and difenzoquat. The analytes were eluted with tetramethylammonium ion solution [115]. Higher concentrations of dissolved compounds may also be determined after extraction as ion pairs with SPE [116] or LLE [117]. Debets et al. [118] investigated electrodialytic sample treatment (EDIST) coupled on line with HPLC, a technique which will be further developed for ionic compounds.

Glyphosate and its metabolite AMPA were determined down to 0.02 $\mu g\,l^{-1}$ after direct derivatization in the 25 ml tap water sample with 9-fluorenylmethyl chloroformate and subsequent LLE [21]. A method for environmental water was collaboratively studied and was adopted by the AOAC as the official first action [17]. In this method, the filtered sample is evaporated to dryness and the residue is dissolved in buffered EDTA solution, which is submitted to post column derivatization HPLC. Again, suspended matter must be analysed separately [17]. The two compounds were also extracted from well and river water with a small anion-exchange column [65].

ETU, a key metabolite of dithiocarbamate fungicides, was analysed down to 1 $\mu g\,l^{-1}$ by direct injection of 200 µl of the sample onto a coupled LC-LC system, see below. For more sensitive determination, 100 ml of ground water sample was evaporated to a small volume and then liquid/liquid extracted prior to the column switch HPLC [119]. In the National Pesticide Survey study [25] 50 ml of water sample was poured onto an Extrelut column after adjustment of pH and ionic strength. The ETU was extracted from the water fixed in the column by 400 ml of dichloromethane. Mean recovery was 92% at a spiking level of 90 $\mu g\,l^{-1}$.

7 Cleanup

The need for cleanup of water sample extracts is dependent on many factors such as type of water, concentration levels targeted, final determination step and degree of selectivity of the extraction procedure. In most cases cleanup is not necessary for drinking water. Most cleanup methods for pesticides have been developed for food analysis but may be applied to extracts of water samples as well. The techniques most frequently used are adsorption and gel permeation chromatography (GPC).

Adsorption chromatography on silica, alumina, magnesia and Florisil (a synthetic magnesium silicate) separates mainly according to polarities and may be quite selective depending on which fractionation is applied. Conventionally, 10–30 g of adsorbents were packed in glass columns, and eluted with large volumes of solvent fractions with increasing polarity. With the higher efficiency of the sorbents in minicartridges, the amount of sorbent and eluents can be greatly reduced, and especially so for water samples, where the amount of co-extractives is normally far less than, for instance, in fatty food. Lopez-Avila et al. [120] compared the cleanup of chlorinated hydrocarbons from surface and leachate water with 12.3 g Florisil, using 600 ml of solvent, with a 1 g cartridge of Florisil and 9 ml of solvent. They found the latter superior. Similarily, hexane extracts of non- and semipolar insecticides in surface water were cleaned up on activated Florisil (0.7 g) packed in a Pasteur pipette and eluted with 15% ethyl acetate in hexane [11]. Silica cartridge were used for cleanup of surface water extracts in the commercially available ASPEC (automated sample preparation with extraction columns) system, which was coupled on line with a capillary GC [60]. Elution patterns from silica gel for over 400 pesticides were presented, see below [121]. Humic acids were conveniently removed from C18 acetonitrile eluate by percolating the eluate through a silica gel cartridge [91, 122]. SPE is well suited for immediate cleanup before elution of the targeted compounds [29]. Typical examples are the methods for thiocarbamates in which the C18 extraction column is washed with water and with 30% methanol solution before elution with methanol [123], and warfarin, which was extracted from water on a C18 column, cleaned up with acetonitrile-water (2 : 8) at pH 4.3 and eluted with acetonitrile-water (1 : 1) at pH 7.4 [80].

Column cleanup by HPLC, with fractions taken off line, or by 'heart cutting' on line, gives much greater efficiency and reproducibility, and very narrow retention time windows may be chosen for single compounds. Seiber et al. [89] used a silica column with a 100% hexane → 100% methyl *tert*-butyl ether gradient for the fractionated cleanup of some forty pesticides of widely different polarities. The fraction periods were indicated by four non-pesticide marker substances with high UV absorptivity injected intermittently. The same column was used for at least one year. A similar system was used for cleanup of derivatized glyphosate [124]. A very selective cleanup was achieved when two precolumns were put in series with an HPLC system. The river water sample was extracted by a PRP-1 precolumn. An eluent of pH 1 ionized the triazines, which were selectively trapped by the following cation exchanger [68].

The application of coupled chromatographic techniques, LC-LC and LC-GC, for cleanup of samples by heart cutting was discussed by van Zoonen et al. [64]. The fraction of interest from the first column is transferred via switching valves to the analytical system. This technique is best suited for single analytes and was applied to chloroallyl alcohol (125), ETU (119) and methylisothiocyanate (126) in ground water. By careful examination of the retention behavior of pesticides, column-switching conditions can be adjusted for multiresidue cleanup [64].

Hydrophobic gel permeation chromatography (GPC) provides universal cleanup in pesticide analysis. Specht and Tillkes presented elution patterns from both GPC and silica columns for over 400 different nonpolar, polar and acidic pesticides [114, 121]. In GPC, compounds are separated mainly by molecular weight although hydrophobic interactions are also involved in the polystyrene gels used. When using the setup for 126 pesticides described by Andersson and Ohlin (39; Bio-Beads SX-3, 7 g in 1 × 40 cm column, eluent cyclohexane : dichloromethane 1 : 1), heavier molecules such as lipids, waxes, pigments and humic acids are eluted in the first 18 ml. The next 20 ml fraction contains compounds with molecular weights of 200–500, including the pesticides. Volatile ECD-sensitive esters of acidic herbicides are obtained after derivatization with PFB (pentafluorobenzylbromide) [43, 127]. However, e.g. fatty acids are derivatized as well but can be separated from the herbicide derivatives by GPC [43]. GPC columns last for several months or years. The eluent may be fed by a levelling vessel [114] but is usually fed by a pump, which makes the systems easy to automate [114]. The GPC cleanup is sufficient in many instances but fractionated Florisil cleanup was found to give cleaner extracts of soil samples compared to GPC cleanup for triazines and their metabolites [128], and organophosphates [129].

There are numerous special techniques for selective cleanup. Thus, many organochlorine pesticides such as DDTs and HCHs are stable towards concentrated sulfuric acid, and when a hexane extract is shaken with this acid, most organic interfering compounds are destroyed [130]. A treatment of the extract with sulfuric acid in an ultrasonic bath for 10 min was tested with 84 pesticides, and was compared with Florisil column cleanup. The latter was preferred because too many pesticides were degraded by the acid treatment and interfering breakdown products were produced [131]. However, a gentle mixing with the acid for half a minute, followed by centrifugation, is less harsh but usually still sufficient.

8 Newcomers' Guidelines

The newcomers to the field would probably start with the analysis of non- and semipolar compounds. They may choose either the LLE or SPE approach. Both techniques could be used for drinking water and most ground waters. For surface water it is advisable to start with the LLE technique. Within ISO and EC-countries methods for the determination of semipolar pesticides in water are now being standardized and will probably be published in the mid 1990s. The procedures outlined below are similar to the methods discussed for standardization.

A one liter sample is suitable in LLE. The ion strength may be increased with some sodium chloride (most convenient as a saturated solution), and then the sample is extracted in a funnel with two to three portions of dichloromethane or ethyl acetate. The extract is dried with sodium sulfate, and is then evaporated in a rotavapor to a volume of about 5 ml. The concentrated extract is carefully transferred to a test tube, and blown down to a small volume in a gentle stream of pure nitrogen. If the solvent has to be changed it should be done gradually by repeatedly adding the new (higher boiling) solvent and blowing down to a small volume. Surface waters generally need further cleanup, in which case gel permeation as described in Refs. [39] and [121] will cover a broad range of pesticides. For more selective cleanup silica or Florisil column chromatography or semipreparative HPLC is advised.

If SPE is chosen it seems easier for the newcomer to get reproducible results with commercial cartridges than with disks. As shown in Fig. 6, the performance of the SPE columns from different purchasers can vary appreciably, and every batch must be tested. The adsorbent preferred is C18. The cartridge should be washed successively with 1–5 ml portions of the eluent, methanol and water. The sample, which may be fortified with sodium chloride and/or 1% methanol, is then forced through at a speed of 5–10 ml min^{-1}. The column is dried with pure nitrogen for 15 min to 2 h depending on SPE column used and the need for dry extract in the final determination. The pesticides are eluted, drop by drop, with 3–4 ml of acetone or methanol. The eluate is concentrated to a suitable volume in a test tube as above.

It is strongly advised that Sect. 2 (General Considerations) should be read before setting up any method.

9 Conclusions

There are many different objectives for analysing pesticides in water. Different analytical approaches are needed for the numerous analyte-matrix combinations possible. The objective must govern which analytes and concentration levels should be targeted, and which pre-extraction and extraction procedures should be considered. Future toxicological and ecotoxicological research will indicate which concentration levels are of interest, and to which extent method development should focus more on the more water soluble compounds and metabolites, or on the particle bound and DOM bound residues.

Presently, the most expedient multiresidue procedures seem to be SPE with membrane disks and, for screening, automated SPE on line with HPLC. The sorbents used in these techniques should be supplemented with sorbents more suited for highly polar compounds. These procedures use only a minimum amount of organic solvents and are less time consuming than LLE. LLE may be preferable when particle and DOM bound pesticides are concerned. Any procedure has to be checked by frequent recovery tests on field samples, with a battery of compounds representing the different chemical characteristics of all analytes targeted.

In conclusion, the choice and performance of an extraction procedure is more crucial for the outcome of the analysis than often conceived. Furthermore, Onuska says (2): "The importance of sample preparation cannot be overemphasized because many of the experimental difficulties encountered result from improperly prepared samples presented for highly sophisticated and expensive instrumental analyses". This, together with awareness of the message forwarded by Norris [40] that "while science and art are important elements in pesticide analysis, craftmanship is essential", will pave the way for sample extracts worth putting into expensive equipment.

10 Examples of Full Procedures

In the author's laboratory the following procedures have proved to produce satisfactory results in the long term [127].

10.1 Determination of Non- and Semipolar Ccmpounds in Surface and Drinking Water

– Sample water in a 1 litre glass bottle with teflon-lined lid. Immediately add 30 ml of dichloromethane. Extract the sample within two days.

– Weigh the bottle with water. Pour the sample, including the dichloromethane, into a separating funnel. Reweigh the bottle and calculate the volume of the water (subtract the weight of 30 ml dichloromethane).

– Rinse the bottle with dichloromethane (25 ml) and transfer to the funnel.

– Add saturated sodium chloride solution (50 ml) to the funnel.

– Shake the funnel mechanically for 2 min. Let phases separate and collect the organic phase in a stoppered E-flask fortified with 5–10 ml of sodium sulfate (pretreated at 760 °C for $\geqslant 4$ h).

– Reextract the sample twice with dichloromethane (2×50 ml).

– Collect the dichloromethane extracts in the E-flask. Shake for 2 min to dry the extract. Add more sodium sulfate if it has been saturated with water and forms a cake instead of floating freely.

– Filter the dried extract into a round bottom flask. Rinse the E-flask and the sodium sulfate with dichloromethane (10 ml). Evaporate at 37 °C on a rotavapor down to a little less than 1 ml. Transfer the residue to a 5 ml graduated test tube. Rinse the round bottom flask thoroughly with small portions of dichloromethane and transfer these to the test tube. Adjust the volume to 1.5 ml. Rinse the flask with small portions of cyclohexane and add to the test tube until the total volume is 3.0 ml. Mix.

– Filter the extract through a 0.45 µm membrane filter. Inject 1 ml of the extract onto a gel permeation column (Biorad SX-3, 10 mm i.d. $\times$ 40 cm, in dichloromethane : cyclohexane 1 : 1; Ref. [39]).

– Elute with dichloromethane : cyclohexane 1 : 1, at a rate of 1 ml min^{-1}. Discharge the first 17–19 ml (the breaking point, which is individual for each column, is where the ethion and diazinon have been found to lose 5% in the first fraction). Collect the following 20 ml in round bottom flask. Rinse the column with 10 ml eluent before injection of next sample.

– Evaporate eluate to about 1 ml. Add 10 ml of cyclohexane and evaporate to about 1 ml. Transfer to graduated test tube. Rinse flask thoroughly with portions of cyclohexane : acetone 9 : 1 and transfer to test tube. Adjust volume to 2.0 ml.

– Screen the extract on GCs equipped with electron capture and nitrogen-phosphorus detectors and nonpolar and semipolar capillary columns, using temperature programming. Verify preferably with capillary GC with mass selective detector.

Comments: For drinking water the gel permeation cleanup is omitted.

Dichloromethane is now being restricted due to environmental considerations. We therefore consider to change the extracting solvent to ethylacetate.

Results are included in Table 3 indicated under Sec 10.1.

10.2 Determination of Phenoxyalcanoic Acid Herbicides and Related Compounds

– Sample water in a glass bottle with teflon lined lid. Store sample at + 4–8 °C. Extract the sample within a week.

Table 3. Pesticides targeted in selected residue methods. x = recovery > 69%; (x) = recovery 40–69%; n = recovery < 40%. (For method in Section 10.1, ** = recovery > 50%; * = recovery 25–50%). In cleanup part no recoveries are given. In ref 121 * = eluent fractions given for GPC only; ** = fractions given for silica as well

	Liquid-liquid extraction												Solid phase extraction																						Additional ref	Cleanup	
Reference no	10	48	7	51 + 52	50	56	14	132	43	55	25 (a)	25 (b)	Sec 10.1	Sec 10.2	11	12	85	42	8	20 no DOM	20 with DOM	84	22	122	29	98	27	132	105	38	94	99	70		121 GPC SiO_2	89 HPLC
acephate																																			**	
acifluorfen												x*																								
alachlor			x			(x)	x	x								x	x	x	x	x	x	x						x				x			**	*
aldicarb								(x)																				x		x		x			*	
aldoxycarb (=aldicarb sulfone)								x																				n				n				
aldicarb sulfoxide								n																				n								
aldrin	x	x								x			**			x		x	(x)																**	
alloxidim-sodium																																x				
ametryn			x																x			x									x				**	
aminocarb																							x													
amitrole																															x					
anilazine																																			**	
atraton			x																							x									**	
atrazine			x		x	x	x						**			x	x	x	x	x	(x)	x		x		x				x	x	x			**	*
atrazine-desethyl								x			x		**					(x)				(x)		x		x		x						46 : x		
atrazine-desisopropyl, see simazine-desethyl																																				
atrazine-OH																																		46 : (x)		
azinphos-ethyl					x								**						(x)			x								x					**	
azinphos-methyl													**						x			x								x					**	*
azinphos-methyl oxon																																				*
barban											x												x									x			**	

(Continued)

Table 3 (Continued)

	Liquid-liquid extraction											Solid phase extraction																					Additional ref	Cleanup	
Reference no	10	48	7	51+ 52	50	56	14	132	43	55	25 (a)	25 (b)	App (a)	App (b)	11	12	85	42	8	20 no DOM	20 with DOM	84	22	122	29	98	27	132	105	38	94	99	70		121 GPC SiO_2	89 HPLC
benalaxyl																						x														
benazolin ethyl ester													**																							
benfluralin					x														x																**	
benomyl																							x													
bentazone				x								x		x				x												x		x				
binapacryl													**																						**	
biphenyl																										x									*	
bitertanol													**																						**	
bromacil			x				(x)						**													x				x		x			**	
bromophos										x			**																		(x)				**	
bromophos-ethyl													**																	x					**	
bromopropylate													**																						**	
bromoxynil									x																					x		x			*	
bromuron																															x					
bupirimate													**												x											
butachlor			x*				x																													
butralin					x																															
buturon																										x										
butylate			x				(x)															x	x							x						
camphechlor										x						x																			**	
captafol						x																													**	
captan										n												(x)	x												**	*
carbaryl								x					**				x					x	x					x		x					*	*
carbendazim																							x												*	
carbetamide																								x												
carbofuran						x		x					**				x	x					x	(x)				x		x					*	*
carbofuran-3-hydroxy																																				*
carbofuran-3-keto																																				*

carbofuran-3-keto-phenol									x																					
carbofuron phenol									x																					
carbo-phenothion											**												x							*	
carboxin		x																													
carboxin sulfone							x																	x							
chloramben										x*																					
chlorbromuron			x*																							x				*	
chlordane-alfa	x										**	x	x	x																**	
chlordane-gamma	x										**	x	x																	**	
chlorfenson											**																			**	
chlorfen-vinphos								x			**																			**	
chloridazon					(x)										x				(x)		x	x				x		n		**	
chlormequat																											n				
chlorobenzilate	x*																													**	
2-chloro-2′,6′-diethyl-acetanilide																													61 : x		
chloroneb	x																													**	
chloro-propylate											**																			**	
chlorothalonil	x										*																			**	
chlorotoluron			x*	(x)												x					x	x					x				
chloroxuron																x						x				x	x			**	
chlorpropham		x									**									x						x	x			**	
chlorpyriphos						x					**			x		x	x	(x)	x						x	x				**	*
chlorpyriphos-methyl																x														**	
chlorsulfuron											**																		86 : x, 134 : x		
chlorthal-dimethyl	x															x														**	*
chlorthal-dimethyl acid metab										x																					
cimoxanyl																			x												

Table 3 (Continued)

	Liquid-liquid extraction										Solid phase extraction																							Additional ref	Cleanup	
Reference no	10	48	7	51+ 52	50	56	14	132	43	55	25 (a)	25 (b)	App (a)	App (b)	11	12	85	42	8	20 no DOM	20 with DOM	84	22	122	29	98	27	132	105	38	94	99	70		121 GPC SiO_2	89 HPLC
clopyralid														x																						
coumafuryl								n																				n		x		x				
coumaphos																														x					**	
4-CPA									x																										*	
crimidine																										x										
crufomate																													x						**	*
cyanazine							x	x			x		**				x									x		x		x					**	
cyanofenphos																																		16:x	**	
cyanophos													**																						**	
cycloate			x																(x)											x						
cyfluthrin													**																							
cypermethrin						x							**																					60:x	**	
2,4-D				x					x			x		x																x			x		*	
2,4-D butoxyethyl ester													**																					r136:x		
2,4-D isooctyl ester													**																					r136:x		
2,6-D																																		133:x		
3,4-D																																		133:x		
dalapon												x*																								
2,4-DB				x					x			x																		x			x		*	
DDD-o,p′																			x																**	
DDD-p,p′	x	x					x			x			**					x	x																**	*
DDE-o,p′																			x																**	
DDE-p,p′	x	x					(x)			x			**				x	x	x			n													**	*
DDT-o,p′		x								x			**						x																**	
DDT-p,p′	x	x			x		x			x			**					x	x			n							x	x					**	*
DEF see tributyl phosphorotrithioate																																				
deltamethrin						x							**																					60:x	**	

demeton-S							x																		*	
demeton-S-methyl																									*	
demeton-S-methyl sulphone																									**	
desmetryn					x						**										x				**	
dialifos											*														**	
di-allate						x																			**	
diazinon			x				x				**		x	x	x	(x)	x		x	x					**	*
diazinon-oxon																			x							
dicamba				x				(x)		x		(x)								x		x				
dicamba 5-OH										x																
dichlobenil					x						**						x			x					**	
dichlofenthion																									**	
dichlofluanid									n		**														**	
2,6-dichlorobenzamide ("BAM")											**															
3,5-dichlorobenzoic acid										x																
dichlorprop				x				x		x		x								x		x	x		*	
dichlorvos			x*		(x)						*			(x)						x					**	
diclobutrazol																		x								
diclofop-methyl						x								(x)											**	
dicloran											**														**	
dicofol																									**	
dicrotophos																									**	
dieldrin	x	x			x		x		x		**			x											**	
dienochlor																									**	
2′,6′-diethyl aniline																								61:(x)		
difenoxuron																				x						
difenzoquat																								30:x		
dimefox																									**	
dimethachlor											**														**	
dimethoate					n	x	x		n		**			(x)			(x)		x	x	x	(x)			**	*
dinobuton					x				x		**														**	
dinocap																									**	
dinoseb										x										x		x				
dinoterb																				x		x				

Table 3 (Continued)

	Liquid-liquid extraction										Solid phase extraction																							Additional ref	Cleanup	
Reference no	10	48	7	51 + 52	50	56	14	132	43	55	25 (a)	25 (b)	App (a)	App (b)	11	12	85	42	8	20 no DOM	20 with DOM	84	22	122	29	98	27	132	105	38	94	99	70		121 GPC SiO_2	89 HPLC
diphenamid			x																			x													**	
dipropetryn					x																														**	
disulfoton			x		(x)																						x			x					**	
disulfoton sulfone			x																																**	
disulfoton sulfoxide			x																																**	
ditalimfos																																			**	
diuron				x				x			x		**													x		x		x		x				
DNOC																														x						
endosulfan-alpha	x*	x			x		x			x			**		x			x	x			x													**	*
endosulfan-beta	x*	x			x		x			(x)			**		x				x			x													**	*
endosulfan ether																			x																	
endosulfan sulfate	x												**		x																				**	
endrin	x	x								x			**			x	x		x																**	
endrin aldehyde	x																																			
EPN													**															x							**	
eptam see EPTC																																				
EPTC			x				(x)						*						(x)			x								x						
ethephon																															n					
ethiofencarb													**																	x						
ethion							x						**						x								x		x						**	
ethirimol																									x											
ethofumesate						x							**																							
ethoprophos			x																															16:x	**	*
etridiazole	x																																		**	
etrimfos					x								**																						**	

fenamiphos		x																						x				**	
fenamiphos sulfone									x																				
fenamiphos sulfoxide						x			x													x							
fenarimol		x																x										**	
fenchlorphos											**												x					**	*
fenitrothion				x				(x)			**			x							x		x	x				**	
fenoprop			x				x			x		x												x		x		*	
fenpropathrin																											60: x		
fenpropimorph					x						**					x													
fenson											**																	**	
fenthion											**													x				*	
fenthion sulphone											**																	*	
fenthion sulphoxide											**																	*	
fenuron			n*																	x					x				
fenvalerate											**		x											x				**	
flamprop												x																	
flamprop-isopropyl												x						x											
fluazifop																			x										
fluazifop-p-butyl																			x									**	
fluometuron			x*			x			x													x		x					
fluotrimazole																												**	
fluridone		x																											
fluroxypyr												x																*	
fluroxypyr isooctyl ester												x																**	
flusilazol					x																								
flutriafol					x														x										
folpet					(x)																							**	
fomesafen																			x										
fonofos															x						x							**	
formothion																												**	
fuberidazole																												**	
glyphosate																									n				
HCH-alpha	x							x			**		x				(x)											**	
HCH-beta	x							x			**						x											**	

Table 3 (Continued)

	Liquid-liquid extraction												Solid phase extraction																					Additional ref	Cleanup	
Reference no	10	48	7	51+ 52	50	56	14	132	43	55	25 (a)	25 (b)	App (a)	App (b)	11	12	85	42	8	20 no DOM	20 with DOM	84	22	122	29	98	27	132	105	38	94	99	70		121 GPC SiO_2	89 HPLC
HCH-gamma	x	x			x	x	x			x			**		x	x	x	x	x			x													**	
HCH-delta	x*						(x)						**						(x)																**	
heptachlor	x	x			x					x			**			x			x			n													**	
heptachlor epoxide	x	x								x			**			x	x		x																**	
heptenophos													*						x								x								**	
hexachloro-benzene	x	x			x					x			**			x			x																**	*
hexaconazole																									x											
hexazinone			x										**													x				x					*	
imazalil																																			**	
iodofenphos													**																						**	
ioxynil														x																x					*	
ioxynil isooctyl ester														x																					**	
iprodione						x					n		*																						**	
isocarbamid																																(x)				
isodrin																																			**	
isofenphos													**																						*	
isopropalin																			(x)																**	
isoproturon													**					x						x		x				x						
lenacil													**																		x				**	
linuron				x*	x		(x)	x			x		**									x				x		x		x		x			**	
malaoxon													*																						**	
malathion					x		x						**		x					x	x	x					x		x	x					**	*
mancozeb																																		56: x		
MCPA				x					x					x				x												x			x		*	*
MCPA isooctyl ester														x																						
MCPB				x					x																					x			x		*	
mecarbam													*																						**	
mecoprop				x					x					x				x												x		x	x		*	

mecoprop butoxyethyl ester												x																	
mephosfolan											**																	**	
merphos see tributyl phosphorotrithioite																													
metalaxyl											**								x									**	
metamitron											*				x					x	x					(x)		**	
metazachlor					x						**				x						x					x			
methabenzthiazuron				(x)*	x						**				x						x							**	
methamidophos																(x)									x			**	
methidathion									n		**						x	(x)	x				x					**	*
methidathion oxon																													*
methiocarb								x														x							*
methomyl								(x)														n		x					*
methoxychlor	x	x			x				(x)		**		x		x									x				**	*
metobromuron				x*											x									x				**	
metolachlor			x		x	x	x							x	x	x			x		x			x		x		**	*
metoxuron				n*							**				x						x			x					
metribuzin			x				x				**			x	x	x					x			x				**	
metribuzin DA (=deaminated)								(x)		x												(x)							
metribuzin DADK										x*																			
metsulfuron-methyl																											134:x		
mevinphos			x		n						**													x				**	
mexacarbate								x														x							
MGK-264			x																										
mirex																												**	
molinate			x													x			x					x				*	*
monocrotophos											**																	**	
monolinuron				x*																	x			x		x		**	
monuron				(x)*	n			x														x		x					
naled																												**	

Table 3 (Continued)

	Liquid-liquid extraction													Solid phase extraction																			Additional ref	Cleanup	
Reference no	10	48	7	51 + 52	50	56	14	132	43	55	25 (a)	25 (b)	App (a)	App (b)	11	12	85	42	8	20 no DOM	20 with DOM	84	22	122	29	98	27	132	105	38	94	99	70		121 GPC SiO_2	89 HPLC
1-naphtylacetic acid																																		133:x	*	
naphtyl-thiourea								n																				n								
napropamide			x																																**	
neburon											x																			x					**	
nicotine																																			*	
nitrofen										x																									**	
4-nitrophenol												x*																							*	*
nitrothal-isopropyl					x																														**	
norflurazon			x																							x									**	
omethoate																															x				**	
oxamyl								(x)																				n		x		n			*	
oxychlordane																																			**	
paclobutrazol																									x											
paclobutrazol-ketone																									x											
paraoxon																														x					**	*
paraoxon-methyl			x																																**	
parathion					x	x	x						**				x		x	x	x	x				x	x		x	x					**	*
parathion-methyl					x		x			n			**		x				x	x	x	x				x	x		x	x	x				**	*
pebulate			x																																	
penconazol													**																							
pencycuron																										x										
pendimethalin																		x	(x)	x	x					x				x					**	
penta-chloroaniline													**																						**	
penta-chloranisol																																			**	

pentachlorobenzene		x								**																			**
pentachlorophenol									x*		x														x		x		**
permethrin	x				(x)					**																	x	60 : x	**
phenmedipham										*									x										
phenthoate																						x							**
phorate														x			x								x				*
phosalone										**				(x)										x					**
phosmet																								x					*
phosphamidon										**																		16 : x	**
phoxim																									x		x		**
picloram							(x)		x																				
piperonyl butoxide																													**
piridafenthion																												78 : x	
pirimicarb										**										x					x				**
pirimiphos-ethyl																									x				**
pirimiphos-methyl										**										x					x				**
prochloraz				(x)						**																			
procymidone										**															x				**
profenofos										**				x															**
prometon		x			x												(x)				x								**
prometryn		x*	x		x					**				x			x				x					x			*
pronamide see propyzamide		x																											
propachlor	x				x	(x)				**		x		(x)									x		x		x		**
propanil								x		**							x								x				**
propargite					x					**																			**
propazine		x	x		x								x	x	x	(x)	x				x			x	x	x			**
propham						(x)		x*		*								x					x		x	x			
propiconazole				x						**			x																**
propoxur						(x)				**								x					x		x				**
propyzamide		x				x				**				x									x		x				**
propyzamide metabolite								x																					
prothiofos										**																			**
pyrazophos										**																		78 : x	**
pyrethrins																													**
quinalphos										**				x															**

Table 3 (Continued)

	Liquid-liquid extraction									Solid phase extraction																							Additional ref	Cleanup	
Reference no	10	48	7	51 + 52	50	56	14	132	43	55	25 (a)	25 (b)	App (a)	App (b)	11	12	85	42	8	20 no DOM	20 with DOM	84	22	122	29	98	27	132	105	38	94	99	70		121 GPC SiO_2	89 HPLC
quino-methionate										n			*																						**	
quintozene		x			x					x			**																						**	
rabenzazole																																			**	
resmethrin																																			**	
rotenone								(x)																				(x)		x						
sebuthylazine					x																					x										
sethoxydim																																x				
siduron								x																				x								
simazine			x		x	x	x						**			x		x	(x)	x	(x)	x				x				x		x			**	*
simazine-desethyl (=atrazine DIP)																		(x)				n		(x)		x								46(x)		
simetryn			x																			x													**	
sulfallate																														x					*	
sulfometuron-methyl																																		86:x		
sulfotep																																		16:x	**	
swep											x																									
2,4,5-T				x					x			x		x																x			x		*	
2,3,6-TBA									(x)																											
TCA																															x					
tebuthiuron			x																																	
tecnazene					x								*																						**	
terbacil			x										**																						**	
terbufos			x																																*	
terbuthylazine					x		x						**						x			x	x			x				x					**	
terbuthylazine-desethyl																																		96:x		
terbutryn			x		x								**						x			x				x									*	
tetra-chloroaniline													**																							*

1,2,3,5-tetra-chlorobenzene		x																				**	
tetrachloro-nitrobenzene										**												**	
tetra-chlorvinphos			x																		78 : x	**	
tetradifon										**						x						**	
tetrasul									x	**												**	
thiacetazone																x							
thiometon					(x)																	*	
thionazin										**												**	
thiram																				x			
tolclofos-methyl										**													
tolylfluanid										**												**	
2,4,5-TP see fenoprop																							
toxaphene see camphechlor																							
triadimefon			x							**						x						**	
triadimenol						x				**												**	
tri-allate										**		x										**	
triazophos										**											78 : x	**	
tributyl phosphoro-trithioate														x	(x)							**	*
tributyl phosphoro-trithioite			(x)																			**	
trichlorfon													(x)									**	
1,2,4-trichloro-benzene		x																				**	
trichloronat										**												**	
triclopyr				x																x			
tricyclazole			x																				
trietazine													x			x						**	
trifluralin	x				x	(x)	x		x	**	x		(x)	x	(x)	(x)	x		x			**	*
vamidothion													n									*	
vernolate			x																		72 : x		
vinclozolin						x			n	*												**	
warfarin								x										x	x	x	80 : x		

* for ref 25 means that these pesticides were presented in ref 45 in addition to those mentioned in ref 25. Figures for non stored samples.
* for ref 7: these pesticides were presented in ref 25 in addition to those mentioned in ref 7.

– Weigh 200 ml of sample into an E-flask. Add 4–5 potassium hydroxid pellets to give a pH of 12. Leave over night or heat in a pressure cooker for 1 h.

– Acidify sample to pH 1–2 with concentrated phosphoric acid (about 5 ml).

– Pour the sample into a separating funnel. Rinse the E-flask with dichloromethane (50 ml) and transfer to the funnel. Shake mechanically for 2 min. Let the phases separate.

– Draw the dichloromethane phase through a plug of glass wool into a round bottom flask.

– Reextract the sample with dichloromethane (50 ml). Evaporate on a rotavapor just to dryness.

– Dissolve in 0.05 M phosphate buffer of pH 8 (4 ml) and transfer to 10–20 ml teflon capped test tube. Add a solution of tetrabutylammonium ion, 0.15 M tetrahexylammonium ion, 0.015 M, prepared in phosphate buffer and titrated to pH 8 (150 μl). Add pentafluorobenzyl bromide (PFB), 0.1% in dichloromethane (2.0 ml). Shake mechanically for 20 min.

– Transfer 1.0 ml dichloromethane phase to a GC glass vial. Evaporate cautiously, preferable by leaving over night in draft hood. Dissolve in 1.5 ml of cyclohexane.

– Analyse by capillary GC with mass selective detector.

Comments: Tetrabutylammonium ion, 0.15 M, can be used alone for phenoxy acids and bentazone. Other compounds may give poor yield.

The PFB derivatives may be detected very well by an electron capture detector. However, fatty acids and other compounds are derivatized as well and may give rise to interferences, and a number of capillary columns with different polaritites may be needed for verifying finds. In MSD the PFB mass fragment (181 m/e) is not selective and should not be used.

Results are included in Table 3 indicated under Sec 10.2.

11 References

1 Barceló D (1991) Analyst 116: 681
2 Onuska FI (1989) J High Resolut Cromatogr 12: 4
3 Namiesnik J, Górecki T, Biziuk M (1990) Anal Chim Acta 237: 1
4 Eadie BJ, Morehead NR, Landrum PF (1990) Chemosphere 20: 161
5 Hadfield ST, Sadler JK, Bolygo E, Hill IR (1992) Pestic Sci 34: 207
6 Sigleo AC, Means JC (1990) Rev Environ Contam Tox 112: 123
7 Edgell KW, Jenkins EL, Lopez-Avila V, Longbottom JE (1991) J Assoc Off Anal Chem 74: 295
8 Davì LM, Baldi M, Penazzi L, Liboni M (1992) Pestic Sci 35: 63
9 Poole SK, Dean TA, Oudsema JW, Poole CF (1990) Anal Chim Acta 236: 3
10 Lopez-Avila V, Wesselman R, Edgell K (1990) J Assoc Off Anal Chem 73: 276
11 Hinckley DA, Bidleman TF (1989) Environ Sci Technol 23: 995

12 Method 525.1 Determination of Organic Compounds in Drinking Water by Liquid-Solid Extraction and Capillary Column Gas Chromatography/Mass Spectrometry (Revision 2.2), Environmental Monitoring Systems Laboratory, Office of Research and Development, U.S. Environmental Protection Agency, Cincinnati, OH USA 45268. Available from: NTIS, 5285 Port Royal, Springfield, VA USA 22161
13 House WA, Ziqing Ou (1992) Chemosphere 24: 819
14 Foster GD, Foreman WT, Gates PM (1991) J Agric Food Chem 39: 1618
15 Marvin CH, Brindle ID, Hall CD, Chiba M (1990) Anal Chem 62: 1495
16 Borburgh HJ, Hammers WE (1992) Tox Environ Chem 35: 79
17 Oppenhuizen ME, Cowell JE (1991) J Assoc Off Anal Chem 74: 317
18 Wells DE, Kelly AG (1991) Mikrochim Acta (Wien) 111: 23
19 Liska I, Krupcik J, Leclercq PA (1989) J High Resolut Cromatogr 12: 577
20 Johnson WE, Fendinger NJ, Plimmer JR (1991) Anal Chem 63: 1510
21 Gauch R, Leuenberger U, Müller U (1989) Z Lebensm Unters Forsch 188: 36
22 Marvin CH, Brindle ID, Singh RP, Hall CD, Chiba M (1990) J Chromatogr 518: 242
23 Simmons ND (1991) Brighton Crop Prot Conf – Weeds, Vol. 3, p1259
24 Landrum PF, Nihart SR, Eadie BJ, Gardner WS (1984) Environ Sci Technol 18: 187
25 Munch DJ, Graves RL, Maxey RA, Engel TM (1990) Environ Sci Technol 24: 1446
26 Schuette SA, Smith RG, Holden LR, Graham JA (1990) Anal Chim Acta 236: 141
27 Mañes Vinuesa J, Moltó Cortés JC, Igualada Cañas C, Font Pérez G (1989) J Chromatogr 472: 365
28 Reupert R, Plöger E, Brausen G (1990) Hewlett-Packard Publ no 12-5952-2229
29 Bolygó E, Atreya NC (1991) Fresenius J Anal Chem 339: 423
30 Ahmad I (1983) J Environ Sci Health B18: 207
31 Fielding M, Gibby S, Moore K (1991) In: Angeletti G, Björseth A (eds) Organic micropollutants in the aquatic environment: proceedings of the 6th European symposium, 1990. Kluwer, Dordrecht, p 142
32 Von Böer G, Schlett C, Thier H-P (1990) Z Wasser-Abwasser-Forsch 23: 220
33 Åkerblom M (1985) J Chromatogr 319: 427
34 Dedek W, Weil L, Feistel L (1992) Vom Wasser 78: 155
35 Leggett DC, Jenkins TF, Miyares PH (1990) Anal Chem 62: 1355
36 Baumgarten D, Pfrang I (1989) GIT Fachz Lab 9: 815
37 Crepeau KL, Walker G, Winterlin W (1991) Bull Environ Contam Toxicol 46: 512
38 Di Corcia A, Marchetti M (1992) Environ Sci Technol 26: 66
39 Andersson A, Ohlin B (1986) Vår Föda 38 Suppl 2: 79
40 Norris LA (1986) Weed Sci 34: 485
41 Weil L, Niessner R (1991) Nachr Chem Tech Lab 39: 1277
42 Scharf J, Wiesiollek R, Bächmann K (1992) Fresenius J Anal Chem 342: 813
43 Lee HB, Peart TE, Carron JM, Tse H (1991) J Assoc Off Anal Chem 74: 835
44 Vogelgesang J (1991) Fresenius J Anal Chem 340: 384
45 Munch DJ, Frebis CP (1992) Environ Sci Technol 26: 921
46 Durand G, Barceló D (1989) Tox Environ Chem 25: 1
47 Baumgarten D, Pfrang I (1989) GIT Supplement 3: 80
48 Geissler A, Schöler HF (1991) Chemosphere 23: 1029
49 Harrington-Fowler L (1991) American Laboratory August 1991: 39
50 Schoeler HF, Brodesser J (1988) In: Angeletti G, Björseth A (eds) Organic micropollutants in the aquatic environment: proceedings of the 5th European symposium, 1987. Kluwer, Dordrecht
51 Färber H, Peldszus S, Schöler HF (1991) Vom Wasser 76: 13
52 Färber H, Schöler HF (1991) Vom Wasser 77: 249
53 Van Hoof F, Ackermans P, Wauters M (1991) In: Angeletti G, Björseth A (eds) Organic micropollutants in the aquatic environment: proceedings of the 6th European symposium, 1990. Kluwer, Dordrecht, p 315

54 Dix KD, Fritz JS (1990) Anal Chim Acta 236: 43
55 Hemmerling C, Risto C, Augustyniak B, Jenner K (1991) Die Nahrung 35: 711
56 Legrand MF, Costentin E, Bruchet A (1991) Environ Technol 12: 985
57 Foster GD, Rogerson PF (1990) Int J Environ Anal Chem 41: 105
58 Gregor DJ (1990) In: Kurtz DA (ed) Long range transport of pesticides. Lewis, Chelsea, p 373
59 Clement RE, Suter SA, Tosine HM (1989) Chemosphere 18: 133
60 van der Hoff GR, Gort SM, Baumann RA, van Zoonen P, Brinkman UAT (1991) J High Resolut Cromatogr 14: 465
61 Potter TL, Carpenter T, Putnam R, Reddy K, Clark JM (1991) J Agric Food Chem 39: 2184
62 Goosens EC, Bunschoten RG, Engelen V, de Jong D, van den Berg JHM (1990) J High Resolut Cromatogr 13: 438
63 Farran A, Cortina JL, De Pablo J, Barceló D (1990) Anal Chim Acta 234: 119
64 van Zoonen P, Hogendoorn EA, van der Hoff GR, Baumann RA (1992) Trends Anal Chem 11: 11
65 Wigfield YY, Lanouette M (1990) Anal Chim Acta 233: 311
66 Pack DE (1967) In: Zweig G (ed) Analytical methods for pesticides, plant growth regulators and food additives, vol. 5. Academic Press, New York, p 397 and p 473
67 Di Corcia A, Marchetti M (1991) Anal Chem 63: 580
68 Coquart V, Hennion M-C (1991) J Chromatogr 553: 329
69 Battista M, Di Corcia A, Marchetti M (1989) Anal Chem 61: 935
70 Di Corcia A, Marchetti M, Samperi R (1989) Anal Chem 61: 1363
71 Hagenguth M, München (1992) personal communication
72 Hagen DF, Markell CG, Schmitt GA (1990) Anal Chim Acta 236: 157
73 Brouwer ER, Lingeman H, Brinkman UAT (1990) Chromatographia 29: 415
74 Brouwer ER, Van Iperen DJ, Liska I, Lingeman H, Brinkman UAT (1992) Int J Environ Anal Chem 47: 257
75 Yoo LJ, Shen Y, Fitzsimmons S (1992) Proc Water Qual Technol Conf 1991, Pt 2, 913
76 Thoma JJ, Kraut A, George JE, Day RS (1992) Proc Water Qual Technol Conf 1991, Pt 2, p 851
77 Junk GA, Avery MJ, Richard JJ (1988) Anal Chem 60: 1347
78 Moltó JC, Picó Y, Font G, Mañes J (1991) J Chromatogr 555: 137
79 Thier H-P, Zeumer H (eds) (1991) Rückstandsanalytik von Pflanzenschutzmitteln/DFG. VCH Verlagsgesellschaft, Weinheim
80 Dalbacke J, Dahlquist I, Persson C (1990) J Chromatogr 507: 381
81 Nash RG (1990) J Assoc Off Anal Chem 73: 438
82 Mangani F, Crescentini G, Palma P, Bruner F (1988) J Chromatogr 452: 527
83 Brooks MW, Jenkins J, Jimenez M, Quinn T, Clark JM (1989) Analyst 114: 405
84 Benfenati E, Tremolada P, Chiappetta L, Frassanito R, Bassi G, Di Toro N, Fanelli R, Stella G (1990) Chemosphere 21: 1411
85 Junk GA, Richard JJ (1988) Anal Chem 60: 451
86 Howard AL, Taylor LT (1992) J Chrom Sci 30: 374
87 Ho JS, Tang PH (1992) J Chrom Sci 30: 344
88 Alberti J, Willemsen H-G (1989) Forum Städte-Hygiene 40: 118
89 Seiber JN, Glotfelty DE, Lucas AD, McChesney MM, Sagebiel JC, Wehner TA (1990) Arch Environ Contam Toxicol 19: 583
90 Mattern GC, Louis JB, Rosen JD (1991) J Assoc Off Anal Chem 74: 982
91 Geerdink RB, Van Balkom CAA, Brouwer H-J (1989) J Chromatogr 481: 275
92 Subra P, Hennion M-C, Rosset R, Frei RW (1989) Intern J Environ Anal Chem 37: 45
93 Subra P, Hennion M-C, Rosset R, Frei RW (1988) J Chromatogr 456: 121

94 Dedek W, Wenzel K-D, Luft F, Oberländer H, Mothes B (1987) Fresenius Z Anal Chem 328: 484
95 Dedek W, Wenzel K-D, Oberländer H, Mothes B, Männig J (1991) Fresenius J Anal Chem 339: 201
96 Ramsteiner KA (1989) J Chromatogr 465: 410
97 Marvin CH, Brindle ID, Hall CD, Chiba M (1991) J Chromatogr 555: 147
98 Reupert R, Zube I, Plöger E (1992) LC-GC Intl 5(6): 43
99 Liska I, Brouwer ER, Ostheimer AGL, Lingeman H, Brinkman UAT, Geerdink RB, Mulder WH (1992) Int J Environ Anal Chem 47: 267
100 Kwakman PJM, Vreuls JJ, Brinkman UAT, Ghijsen RT (1992) Chromatographia 34: 41
101 Arthur CL, Potter DW, Buchholz KD, Motlagh S, Pawliszyn J (1992) LC-GC 5(10): 8
102 Louch D, Motlagh S, Pawliszyn J (1992) Anal Chem 64: 1187
103 DiGiano FA, Elliot D, Leith D (1988) Environ Sci Technol 22: 1365
104 Kaiser RE, Rieder R (1989) J Chromatogr 477: 49
105 LeBel GL, Williams DT, Griffith G, Benoit FM (1979) J Assoc Off Anal Chem 62: 241
106 Green DR, Stull JK, Heesen TC (1986) Mar Pollut Bull 17: 324
107 Ahnoff M, Josefsson B (1974) Anal Chem 46: 658
108 Gómez-Belinchón JI, Grimalt JO, Albaigés J (1988) Environ Sci Technol 22: 677
109 Strachan WMJ, Huneault H (1984) Environ Sci Technol 18: 127
110 Nilvé G, Audunsson G, Jönsson JÅ (1989) J Chromatogr 471: 151
111 Nilvé G, Stebbins R (1991) Chromatographia 32: 269
112 Nilvé G (1992) Sample pretreatment methods for determination of acidic herbicides in water, with special emphasis on supported liquid membranes Thesis, Lund University, Lund, Sweden
113 Mathiasson L, Nilvé G, Ulén B (1991) Intern J Environ Anal Chem 45: 117
114 Thier H-P, Zeumer H (eds) (1987) Manual of pesticide residue analysis/DFG. VCH Verlagsgesellschaft, Weinheim
115 Reichert JK, Lochtman J, Huschens O (1991) Gewässerschutz, Wasser, Abwasser 125: 568
116 Gill R, Qua SC, Moffat AC (1983) J Chromatogr 255: 483
117 Åkerblom M (1990) Bull Environ Contam Toxicol 45: 157
118 Debets AJJ, Hupe K-P, Kok WT, Brinkman UAT (1992) J Chromatogr 600: 163
119 Hogendoorn EA, van Zoonen P, Brinkman UAT (1991) Chromatographia 31: 285
120 Lopez-Avila V, Dodhiwala NS, Milanes J, Beckert WF (1989) J Assoc Off Anal Chem 72: 593
121 Specht W, Tillkes M (1985) Fresenius Z Anal Chem 322: 443
122 de la Vigne U, Jänchen DE, Weber WH (1991) J Chromatogr 553: 489
123 Cubero Ferradal F, Rodriguez Procopio J, Hernandez Hernandez L (1992) Fresenius J Anal Chem 343: 382
124 Seiber JN, McChesney MM, Kon R, Leavitt RA (1984) J Agric Food Chem 32: 678
125 Hogendoorn EA, de Jong APJM, van Zoonen P, Brinkman UAT (1990) J Chromatogr 511: 243
126 Hogendoorn EA, Brinkman UAT, van Zoonen P (1991) Poster presented at: 3rd Workshop on chemistry and fate of modern pesticides, 4–6 Sept 1991, Int Ass Env Anal Chem
127 Åkerblom M, Thorén L, Staffas A (1990) Vår Föda 42: 236
128 Durand G, Forteza R, Barceló D (1989) Chromatographia 28: 597
129 Barceló D, Solé M, Durand G, Albaigés J (1991) Fresenius J Anal Chem 339: 676
130 Murphy PG (1972) J Assoc Off Anal Chem 55: 1360
131 Bernal JL, Del Nozal MJ, Jiménez JJ (1992) J Chromatogr 607: 303

132 Bellar TA, Budde WL (1988) Anal Chem 60: 2076
133 Ahmed AW, Mallet VN, Bertrand MJ (1989) J Assoc Off Anal Chem 72: 365
134 Cotterill EG (1992) Pestic Sci 34: 291
135 Åkerblom M (1990) Pesticide residue analysis with special reference to ion pairing techniques. Thesis, Swedish Univ of Agr Sci, Uppsala, Sweden
136 Bogus ER, Watschke TL, Mumma RO (1990) J Agric Food Chem 38: 142

Capillary GC with Selective Detectors (ECD, NPD, FPD)

Patrick T. Holland

Horticulture and Food Research Institute of New Zealand,
Ruakura Research Centre, Private Bag, Hamilton, New Zealand

Theory and principles of high resolution gas chromatography (HRGC) using fused silica open tubular columns (FSOT) are discussed. This includes type descriptions and the functions of columns, carrier gases, injection devices, detectors, and complex multi channel systems.

A short excursion is made to on-line liquid chromatography/HRGC combination as a logic development step in the sample pre-prepared introduction technique to gas chromatography.

General recommendations for handling the most efficient gas chromatographic parameters are outlined, respective problems are discussed and guidelines to overcome these problems and to optimise the gas chromatographic procedures for trace pesticide analysis in water are communicated.

The chapter is completed by practical directions for quantitation, confirmation and reporting the data and is illustrated by several figures on chromatography of real water samples.

Chemistry of Plant Protection Vol. 11

1 Introduction

Gas chromatography (GC) is the basis for the majority of primary screening methods for pesticides in food and environmental samples such as water. The high resolving power and the various highly sensitive and selective GC detectors have made practical residue determinations for a wide range of pesticides at μg/litre (ppb) levels and below. Despite these advantages the current European limits for pesticides in drinking water are not easy to enforce with current technology. The scope of the problem can be summarised in the calculation that limits of detection (LODs) for individual analytes of 0.05 ppb correspond to extraction of a 1 litre water sample to give a final extract volume of 0.5 ml which must yield a detectable peak in the chromatographic system for 100 pg analyte per μl of extract injected. Only carefully optimised and maintained GC systems can routinely produce reliable data for a wide range of pesticides at these levels.

Open tubular (capillary) columns with wall-coated stationary phases increase the power of GC methods. Their higher resolution enables separation of more individual pesticides from each other and from co-extractives. Detection limits are lowered for the narrow peaks. Fused silica open tubular (FSOT) column technology has become available since the early 1980s. The high resolution, inertness and ruggedness of FSOT columns with bonded stationary phases have greatly increased the reliability and scope of residue determination by GC. Residue methods using HRGC with FSOT columns can give results for a wider range of pesticides, including more polar compounds, than were possible with earlier packed column technology. This is of particular importance for analysis of water as pesticides in the medium polarity range tend to be those most readily transported to water by run-off or leaching.

GC can be applied to many pesticide classes including virtually all insecticides and the majority of herbicides and fungicides. However there are a number of pesticide classes which are not amenable to direct GC determination in their parent form due to involatility or thermal instability. These include (with common examples):

1. Acidic herbicides:	phenoxy acids, benzoic acids, propionic acids, triclopyr, picloram.
2. Polar herbicides:	paraquat, diquat, glyfosate.
3. Phenylurea herbicides (some):	diuron, neburon.
4. Sulfonylurea herbicides:	chlorsulfuron, metsulfuron.
5. Oxime-carbamate insecticides:	aldicarb, oxamyl.
6. Benzimidazole fungicides:	benomyl, MBC.
7. EBDC fungicides:	maneb, mancozeb, zineb.
8. Polar fungicides:	triforine, phosethyl-Al.

Many of these classes can be determined by GC following simple derivatisations such as esterification or alkylation.

Modern commercial GC instruments represent a relatively stable and mature technology capable of achieving high standards of reliability and reproducibility. Optimisation of the injection, separation and detection systems can produce analytical methods of great power. Thus multi-methods can now reliably screen for over 100 pesticides in a water extract using a single GC run of less than 40 min. However GC for residue determination has not yet become an entirely routine technique. This is principally due to the difficulty of maintaining good GC performance for often labile pesticides at pg levels in systems undergoing subtle changes with time due to the effects of heat, oxidation and build up of involatile co-extractives. Also the selective detectors used for residue analysis are less stable and trouble free than the flame ionisation detector (FID) commonly used in other GC applications.

This review summarises the principles of high resolution gas-liquid chromatography (HRGC), with an emphasis on those aspects important to residue analysis using narrow bore FSOT columns with selective detectors, and makes practical recommendations for the application of HRGC to screening of water samples for pesticide residues. Progress in direct coupling liquid chromatographic techniques for sample extraction or cleanup to HRGC is reviewed. Analyses for volatile pesticides or fumigants using specialised GC techniques such as head-space or purge-and-trap analysis with wide-bore columns (Megabore, 0.53 mm i.d. FSOT) are not covered. Although megabore columns are an effective upgrade for older packed column instruments [1] and have advantages for analysis of highly volatile compounds, they cannot achieve the full benefits of HRGC for most residue work.

2 Separation Principles

The most common columns for HRGC are FSOT with i.d.s of 0.1 to 0.32 mm and lengths of 10 to 50 metres. A fundamental requirement for effective use of GC in residue analysis is the separation of the various pesticides of interest from each other and from coextractives. The following equations are valid for Gaussian peaks:

$$\text{Resolution, } R_{12} = 1.18\frac{(t_2 - t_1)}{(W_1 + W_2)} \tag{1}$$

$$R_{12} = \frac{\sqrt{N}}{4}\frac{(a_{12} - 1)}{a_{12}}\frac{k_2}{(1 + k_2)} \tag{2}$$

$$\text{Number of theoretical plates, } N = 5.54\, t_2^2/W_2^2 \tag{3}$$

$$\text{Column selectivity (Corrected relative retention), } a_{12} = \frac{t_2 - t_d}{t_1 - t_d} \tag{4}$$

$$\text{Partition ratio, } k_x = (t_x - t_d)/t_d \qquad (5)$$

where: W_x = peak width at half-height of component x
t_x = retention time of component x
t_d = dead time (retention time of inert gas)

Resolution, defined by the degree of separation of 2 components (Eq. 1), is determined by 3 factors (Eq. 2):

1. Efficiency of the column i.e. the narrowness of the peaks as governed by the plate count (Eq. 3).
2. Stationary phase selectivity to different chemical characteristics of the components (Eq. 4). To achieve a resolution of two components a_{12} must be greater than 1. Resolution of a pair of components is enhanced by a phase which interacts more with one component than the other. This may be due to boiling point differences in or other chemical characteristics.
3. Degree of retention of the component relative to an unretained substance as expressed by the partition ratio or capacity factor, k (Eq. 5). Resolution is enhanced by longer retention times until k reaches about 10. Substances which are not retained cannot be separated.

Note the equations are derived for isothermal conditions. The same factors operate in temperature programmed separations but cannot be so conveniently expressed in mathematical form. The partition ratio has an inverse exponential relationship with temperature and so column temperature is a key determinant of component retention. The partition ratio is approximately halved by a 20 °C increase in column temperature.

Resolution can be expressed according to Kaiser [2] as the separation number or Trennzahl, the number of peaks that can be fitted between two consecutive homologues differing by a methylene unit which have resolution R:

$$\text{Trennzahl} \quad TZ = (R/1.77) - 1 \qquad (6)$$

This measure of column separating power has the advantage of also being applicable to temperature programmed separations and is rather easier to visualise. The interaction of column efficiency and retention on resolution of components and Trennzahl is illustrated schematically in Fig. 1.

2.1 Column Efficiency

Efficiency can be expressed as the number of theoretical plates generated per unit length of column or its inverse, the height of column, H, equivalent to a theoretical plate. The Van Deemter equation as reformulated by Golay [3] provides an accurate description of column efficiency in HRGC. For thin stationary phase films and columns operating with modest pressure drops it may be simplified to:

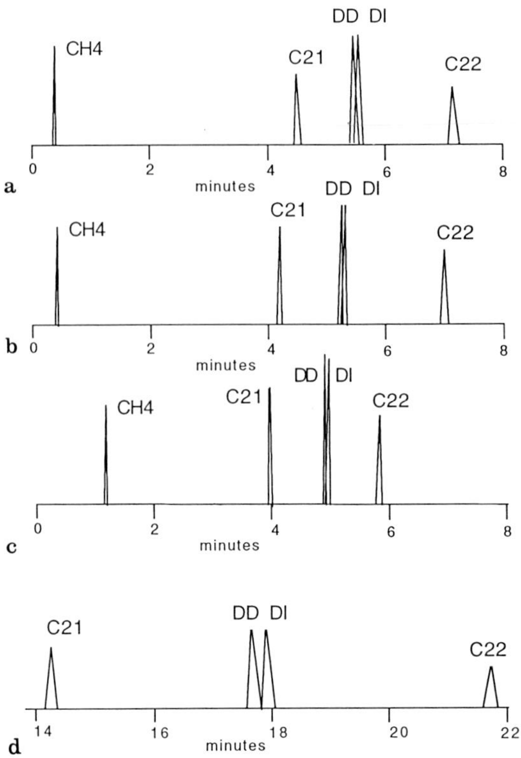

Fig. 1a–d. Simulated isothermal HRGC of C_{21} + C_{22} *n*-alkanes, 4, 4′-DDE (DD) and dieldrin (DI) on 5% phenyl 95% methyl silicone FSOT columns
N = number of theoretical plates
k = partition ration for dieldrin
R = resolution for 4, 4′-DDE/dieldrin
TZ = Trennzahl (separation number)
a) 10 m × 0.53 mm i.d., 1 μm film, 200 °C
N = 15000 k = 12 R = 0.41 TZ = 12.
b) 10 m × 0.32 mm i.d., 0.5 μm film, 200 °C
N = 30000 k = 12 R = 0.58 TZ = 17.
c) 30 m × 0.32 mm i.d., 0.5 μm film, 230 °C
N = 90000 k = 3 R = 0.81 TZ = 23.
d) 30 m × 0.32 mm i.d., 0.5 μm film, 200 °C
N = 90000 k = 12 R = 1.0 TZ = 29.

Height equivalent to a theoretical plate

$$H = N/L$$
$$= 2D_m/U + R_m U \qquad (7)$$

where: L = length of column

U = average linear velocity of the carrier gas
D_m = diffusion coefficient of the component in the carrier gas
R_m = resistance to mass transfer in the carrier gas

At low carrier velocities the first term is important which describes the longitudinal diffusion of component bands in the gas phase, while at medium to high carrier velocities the second term dominates which covers transfer of components between the gas and liquid phases. Resistance to mass transfer has a complex dependance on the partition ratio or capacity factor, and also follows the square of the internal diameter of the column and the inverse of D_m. The overall relationship of H to carrier gas velocity is parabolic with steeper curves for nitrogen than helium or hydrogen carrier gases. The minimum plate height is reached at low carrier gas velocities which will not be the optimum for speed of separation. Reformulations of the Golay equation are possible to optimise the carrier gas velocity and column length to achieve a required resolution in the minimum time [4] and also to maximise detectability [5].

2.2 Stationary Phases

Although a wide variety of phases are available for HRGC, selectivity is often not as important as for low resolution packed columns. Secondary characteristics of particular phases such as stability and low bleed are often more valuable than polarity for residue analysis. Highly polar phases were required for particular packed column applications such as the separation of dieldrin from p,p-DDE which can be accomplished by HRGC on low polarity phases.

The most suitable FSOT columns for trace analysis are manufactured with polymer phases based on linear methyl silicones. These are cross-linked and chemically bonded in-situ to the residual hydroxyl groups on the silica column wall. This makes the film more stable to bleed and oxidation and contributes to deactivation of the column wall. The relatively complex processes used to prepare the fused silica surface, coat the phase and carry out the bonding/cross-linking are now largely carried out by commercial companies. The art of laboratory preparation of columns is being lost with an attendant large increase in the cost over packed columns. However, the commercial columns now available have very high performance specifications (resolution, low bleed, inertness and uniformity) which largely render them indispensable to busy analytical laboratories.

Alteration of selectivity from principally boiling point in bonded silicone phases is achieved by substitution of a proportion of the methyl groups with more polar moieties. The characteristics of the most common phases used in HRGC of pesticides are summarised in Table 1. Formal descriptors have been developed for the various types of interaction of stationary phase with analytes. The McReynold polarity scale classifies packed column phases on the basis of the relative retentions of 5 probe compounds. This system has been applied to

Table 1. Silicone polymer bonded phases for HRGC of pesticides

Phase	Commercial Equivalents	Selectivity
100% methyl silicone	BP-1, DB-1, HP-1, CP-Sil-5, RSL150, SPB1, Ultra 1	Non-polar – boiling point
95% methyl + 5%phenyl	BP-5, CP-Sil-8, DB-5, HP-5, RSL-200, SE-54, SPB-5, Ultra 2	Low-polarity – mainly boiling point
50% methyl + 50% phenyl	BP-17, DB-17, HP-17, OV-17, RSL-300, Rtx-50, SP-2250	Medium polarity – aromatics, unsat'n.
50% methyl + 50% trifluoropropyl	DB-210, RSL-400, SP-2401	Medium-high polarity – carbonyls
86% methyl + 14% phenylcyanopropyl	BP-10, CP-Sil-19, DB-1701, OV-1701, Rtx-1701, SPB-7	Medium-high polarity – hydroxyls, unsat'n.
50% methyl + 50% phenylcyanopropyl	BP-225, CP-Sil-43, DB-225, HP-225, RSL-500, SP-2330	High polarity – hydroxyls, unsat'n.

FSOT columns [6]. However, the wide range of chemical classes and individual functional groups represented amongst pesticides has led to a more empirical approach to HRGC column selection for residue analysis. The least polar phase adequately separating the required components will generally be the most satisfactory because of the reduced stability and lower coating efficiency of the more polar phases. Where critical pairs of components require separation, consideration of particular functionalities can assist selection of an appropriate phase based on the principle of like-to-like e.g. selective retardation of aromatic compounds on phenyl silicone phases or compounds containing P=O and carbonyl groups on trifluoropropyl phases.

2.3 Selection of Columns

The primary choice of phase for screening of residues by HRGC is a low polarity bonded phase, either 100% methyl silicone or 5% phenyl. For screening of water extracts for low level residues there has been some preference for the later type which seem to have had more reliable low bleed and deactivation characteristics. A more polar column is generally used for confirmatory purposes. The choice of phase is not critical provided good column performance can be maintained at residue levels (pg) and the shifts in relative retention times from the screening column are adequate to give confirmatory data. A limiting factor can be column stability and bleed at the temperatures required for late eluting compounds such as synthetic pyrethroids. Methyl silicones with 50% phenyl or 14% phenylcyanopropyl have been common choices.

Other variables in column selection are the column internal diameter (i.d.), column length and stationary phase film thickness. Resolution is dependent on the square root of column length and approximately the inverse of column i.d. (Eqs 2 and 7). Furthermore, increasing the column length proportionally increases the analysis time at given conditions of column temperature and carrier velocity. It follows that using a smaller i.d. column can be more effective than increasing the column length to achieve a required separation.

The component retention and sample capacity are increased by a thicker stationary phase film. The partition ratio k (Eq. 5) is proportional to the column phase ratio, β (Eq. 8, gas volume: film volume). Thus β and therefore k are unchanged if the column i.d. and film thickness are both halved.

$$\text{Column phase ratio} \quad \beta = \mathrm{ID}/4\mathrm{d} \tag{8}$$

where ID = column i.d.
d = film thickness

The column sample capacity is dependant on β and also increases with approximately the square of the column i.d.. Thus halving both the column i.d. and the film thickness decreases the sample capacity by a factor of 4. Exceeding the sample capacity or overloading the column reduces the resolution for the overloaded components, initially by band broadening followed by a prominent leading edge at higher levels. For a 0.2 mm i.d. column with a 0.25 μm film thickness the sample capacity is approximately 30–60 ng depending on the chemical nature of the compound. This will reduce to 10–15 ng for a 0.1 mm i.d. column with the same β. These limits are adequate for the levels of pesticides encountered in most analyses of waters. However the levels of coextractives can more readilly exceed the column capacity, which may give rise to broader interfering peaks, depending on the selectivity of the detector. Pesticide peaks also may be broadened if they elute close to heavily overloaded components.

The stationary phase film has a secondary effect in providing a higher degree of deactivation of the column wall. Very thin films (less than 0.1 μm) should be avoided for HRGC of labile pesticides. However thick film columns (> 1 μm) are more prone to bleed, have steeper Van Deemter curves and are more retentive, possibly leading to excessive analysis times or the need for high column temperatures.

A good selection of screening column for residue analysis with current HRGC technology is a 25–30 m by 0.2–0.25 mm i.d. FSOT column coated with 0.15–0.33 μm of 5% phenyl/95% methyl silicone bonded phase This will give a separation efficiency of about 90 000 theoretical plates. Narrower bore FSOTs appear an attractive alternative but have not been much used in routine screening for pesticides. A 10 m by 0.1 mm column with 0.1 μm film can supply equivalent separations to a 25 m by 0.2 mm i.d. column with analyses completed in a half to one third the time. The requirements for effective use of narrow bore columns have been described [7, 8]. However, columns with smaller i.d. have reduced sample capacity, higher pressure drops and the extra column contributions to band broadening become more critical. The practical limits, with standard HRGC instruments using optimised injectors, are 10–15 m columns of 0.10 mm i.d..

2.4 Column Carrier Gas Operating Conditions

Although nitrogen carrier gas provides higher resolution at optimal flows, helium or hydrogen are used for most HRGC due to their flatter Van Deemter curves. This allows use of higher carrier velocities without large reduction in resolution, leading to much shorter analysis times. Hydrogen is the most favoured gas in this respect, but it is not suitable for some selective detectors including heated bead nitrogen-phosphorus detectors, which require very stable, low flows of hydrogen. Thus helium is generally used as the carrier gas. The carrier gas velocity should be set to about 1.5–2 times the optimum, i.e. 30–45 cm s^{-1}, to minimise analysis time. Note with pressure regulated gas flows this velocity should be set with the column temperature near the upper end of the operating range. The viscosity of helium is reduced at lower temperatures which results in substantially higher column flows (2.5–3 fold for 80 °C from 280 °C). This is advantageous for sample transfer to the column during injection (see below) although resolution of early eluting components may be somewhat reduced.

2.5 Column Temperature Operating Conditions

The column oven temperature programme is an important determinant of analysis time and separation. The temperature during injection will be set near the boiling point of the solvent dependent on the injection technique chosen (see below), followed by a fast ramp to the initial temperature required to begin the separation. The initial working temperature and the following ramp(s) must be adjusted empirically to optimise resolution for critical pairs of components and to minimise analysis time. The initial and final T can be adjusted using the retention times of the first and last components of interest as measured from the end of the first and last ramp respectively. For the ideal isothermal partition ratios of 8–10 these times should be about 10 min on a 25-m column and the speed of the intermediate ramp(s) should not exceed about 8 °C/min. Faster

Fig. 2a–c. Effect of temperature programme rate on the separation of organochlorine insecticides.
Column: 25 m × 0.32 mm i.d. Ultra-2 0.54 μm film. HSI: 250 °C. Initial programme: 80 °C 1 min 30 °C/min to 175 °C followed by: **a**) 3 °C/min to 290 °C **b**) 4 °C/min to 290 °C **c**) 5 °C/min to 290 °C
Peak identity: 1, α-HCH; 2, β-HCH; 3, γ-HCH; 4, δ-HCH; 5, heptachlor; 6, aldrin; 7, heptachlor epoxide; 8, endosulfan I; 9, dieldrin; 10, 4,4′-DDE; 11, endrin; 12, endosulfan II; 13, 4,4′-DDD; 14, endrin aldehyde; 15, endosulfan sulphate; 16, 4,4′-DDT.
Reproduced with permission from Ref. [9]

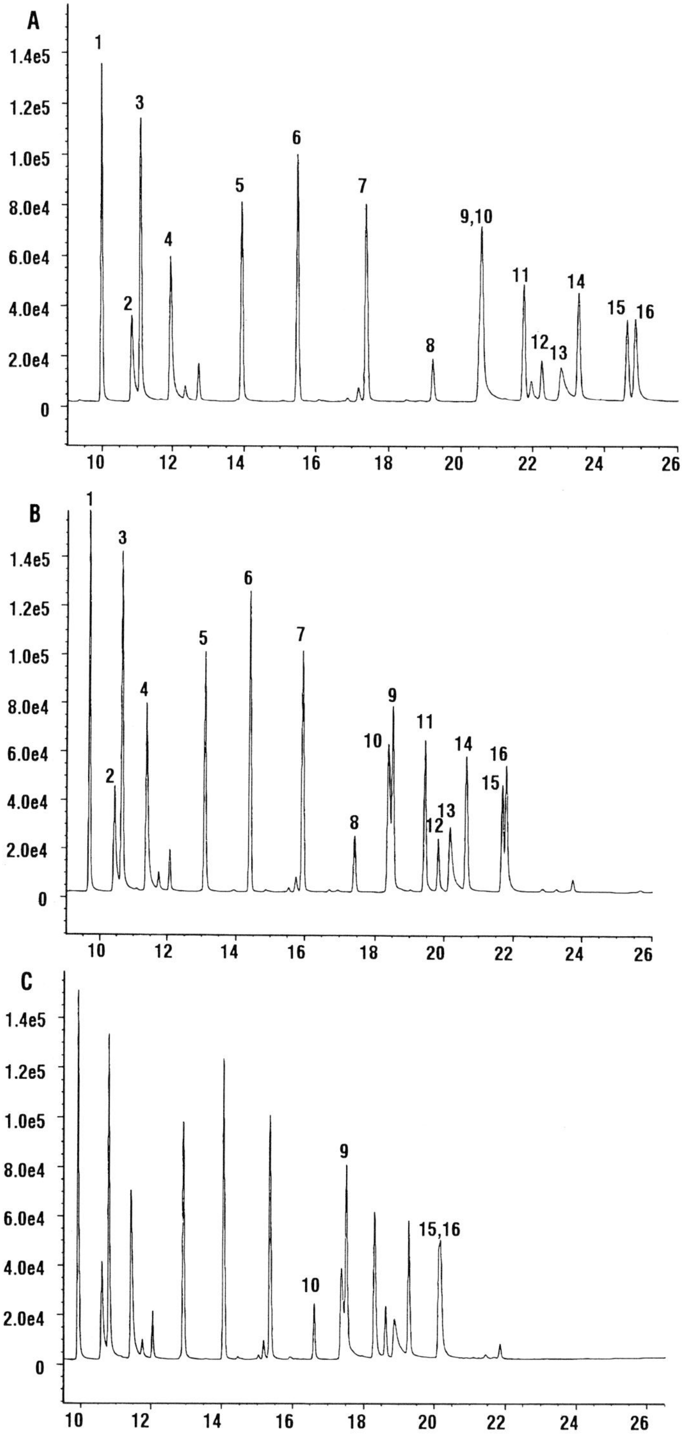
A
1.4e5
1.2e5
1.0e5
8.0e4
6.0e4
4.0e4
2.0e4
0
1 2 3 4 5 6 7 8 9,10 11 12 13 14 15 16
10 12 14 16 18 20 22 24 26
B
1.4e5
1.2e5
1.0e5
8.0e4
6.0e4
4.0e4
2.0e4
0
1 2 3 4 5 6 7 8 10 9 11 12 13 14 15 16
10 12 14 16 18 20 22 24 26
C
1.4e5
1.2e5
1.0e5
8.0e4
6.0e4
4.0e4
2.0e4
0
10 9 15,16
10 12 14 16 18 20 22 24 26

ramps will reduce analysis time at the expense of resolution. A slower ramp may be advantageous in a particular part of the retention region where resolution is critical. An additional ramp after the last component of interest, raising the column temperature quickly to near the operating limit, is useful to remove late eluting coextractives from the column that may interfere with subsequent injections. Where retentions of key pairs of components show different temperature dependencies, the degree of separation may depend on programme rate. Figure 2 shows the separation of a range of organochlorine insecticides on a non-polar column [9]. Note the effect of small changes in the programme rate on the critical separations dieldrin/4,4′-DDE and endosulfan sulfate/4,4′-DDT.

3 Sample Introduction for HRGC

Sample introduction is a critical process for GC of pesticides and other xenobiotic compounds in extracts of water. The usual method is injection of a small aliquot of the sample extract dissolved in a suitable solvent. Sample introduction involves 3 functions:

a) The volatilisation of sample components and transfer into the separating column.
b) The initial separation of solvent from sample.
c) The retention of involatile coextractives.

The quantitative and reproducible accomplishment of these processes for trace analysis is not straightforward, particularly when automated analyses of hundreds of extracts are required for a wide range of compounds many of which are labile. The following discussion is centered on the 3 most common types of injector. There are several variants to each type so only the main principles are outlined. Split injection is not considered in detail as it is of limited utility in trace analysis except for system testing. The direct coupling of liquid chromatography or other sample preparation method to HRGC is a current goal of chromatography research which has already demonstrated considerable potential and is discussed as one of the technologies for the future in routine analysis of water samples for pesticide residues. Each of the introduction techniques depend on thermal trapping or the solvent effect as described by Grob [10, 11] to achieve a focussing of the sample components in a narrow band at the head of the column prior to their volatilisation for separation.

3.1 Cold On-Column Injection (OCI)

Conceptually the simplest injector for HRGC, it also can be the most effective in transfering sample to the column with minimum degradation or discrimination against high boiling components. There is no heated injector zone but some designs incorporate means to cool the valve and column head region during injection using ancillary air or the whole injector may be movable to bring 20–30 cm of column out of the oven. The syringe is introduced directly into the head of the column which is held at, or slightly below, the boiling point of the solvent. For small volumes (1 μl) a fast injection technique minimises discrimination from sample deposition on the needle. For larger volumes a relatively slow injection is used (0.2–0.5 $\mu l\, s^{-1}$) and the solvent creates a flooded zone at the head of the column. The sample components are concentrated in the tail of the solvent film as it evaporates and are left in a narrow band to be volatilised and separated when the column oven temperature is raised [11].

Early versions of the OCI for HRGC used special injector valves to allow the introduction of a fine syringe needle into narrow bore (0.20–0.32 mm i.d.) columns. More recently standard 0.47 mm o.d. syringe needles have been used with standard septum seals to inject into 0.53 mm i.d. deactivated FSOT precolumns. This arose out of the work of Grob showing that a retention gap consisting of 1–10 m of uncoated column is advantageous for the solvent effect, particularly when larger volumes of solvent than 1–2 μl are injected. A wide bore pre-column can be coupled to a narrow bore analytical column thus allowing the use of standard injector components including autosamplers for OCI-HRGC. The extension of this research has been the injection of large volumes of solvent (50–200 μl) using OCI onto long retention gaps. This process has been automated with an HPLC autosampler and rotary valve for HRGC determination of US-EPA method 625 priority pollutants using 50–200μl injections of dilute extracts [12].

3.2 Hot-Splitless Injection (HSI)

HSI was the first injection technique to be perfected for trace analysis using HRGC and remains one of the most useful for pesticide residues. However, the fundamentals of HSI operation are more complex than those of other injection techniques and optimisation is not straightforward. The basic principle is rapid vapourisation of the injected sample in a high temperature injector with the split valve closed and the column temperature below the boiling point of the solvent. The sample components are trapped in narrow band in the cool stationary phase at the head of the column, aided by partial recondensation of the solvent. The split valve is then opened to flush residual solvent from the injector and the column is raised to operating temperature. Involatile coextractives remain in the injector volume and have minimal effect on the column, aided by the split vent being open except during injection (30–90 s).

Early designs were based on split injectors or used modified packed column injectors. The injection chamber is a glass tube (2 mm–4 mm i.d.) with a sliding fit in the injector port. Carrier gas regulation is by adjustment of the column head pressure and the split restrictor valve. A septum purge reduces the effects of septum bleed and flash-back of solvent vapour.

More recent designs use a glass insert which is internally sealed to the injector barrel and contains restrictions at the needle entry and/or the column entry ends. Carrier gas regulation is by control of the total flow through the injector (split + column + septum purge) plus back pressure regulation to control the column head pressure. This design minimises sample contact with metal injector parts.

The extensive studies of Grob and colleagues have established the key requirements for good quantitative performance and high resolution using HSI [10]. Studies on optimisation of operating conditions for several commercial designs have also been published [13–17]. The optimal conditions appear to strongly depend on the detailed construction of each injector, the application of interest and the solvent chosen for the study. However some clear general recommendations applicable to use of HSI for pesticide residue analysis include:

1. The column should be installed so the needle tip reaches within 1–2 cm of the column end to ensure maximum transfer of sample to the column.
2. The injector temperature should be as high as possible to reduce discrimination against high boiling components. This requirement will be a compromise with the need to minimise thermal degradation of sensitive pesticides.
3. The glass injector insert should be thoroughly deactivated and have an internal volume of 0.25–1 ml (see discussion below).
4. The linear flow rate of carrier gas through the injector should be as high as possible so that transfer of sample to the column is rapid. Programming of the column head pressure has been shown to be advantageous for HSI of sensitive pesticides [18]. However the flow rates obtainable from constant pressure control should be adequate if the optimum linear flow is set at high column temperatures (see Sect. 2).
5. The solvent boiling point should be as high as possible, as governed by adequate separation of the solvent and impurities from early eluting pesticides. The lower volume of vapour from high molecular weight, low density solvents minimises the amount of flash back in the injector. Solvents such as iso-octane, cyclohexane and ethyl acetate are suitable for most pesticide work. Additional advantages over more volatile solvents are the higher initial column temperatures and the reduced evaporative losses from prepared solutions.
6. The samples and standards should be in the same solvent and injected volumes should be identical. Retention times for early eluting components are influenced by the solvent in HSI. There can also be a strong influence on response factors. For example Lee et al. [19] showed large differences in FID response factors for polynuclear aromatic hydrocarbons when using HRGC with HSI depending on which solvent the standards were prepared in. This undoubtedly mainly arises from differences in the transfer efficiency of sample components from HSI to column including effects of solvent type on the degree of flash-back and on surface activity in the injector.
7. The initial column temperature should be 10–20 °C below the boiling point

of the solvent to ensure solvent focusing of sample components on the head of the column.

8. For manual injections of 1–2 μl the hot needle technique minimises discrimination. Sample is loaded into the syringe and air drawn into the needle. The needle is put into the injector and after 5 s the plunger is depressed rapidly. The syringe is withdrawn after a further 5 s.
9. Rapid injection is effective for autosamplers with fast injection capability which can insert, inject and withdraw the syringe in less than 0.5 s. A solvent plug may be advantageous for transfer of high boiling pesticides and to reduce carryover [17].
10. The split valve should be reopened 30–60 s after completion of injection. This time may need to be extended to up to 120 s for complete sample transfer when operating with large injector chambers at low flows.
11. The split flow need not be high to clean out the injector. 30 ml min^{-1} is adequate and will save expensive helium over higher flows.
12. Injections of volumes larger than 2 μl are possible but optimal conditions may be more difficult to set. A slow injection will be required (0.5–1 μl s^{-1}) using a precolumn/retention gap of deactivated FSOT (2–10 m depending on i.d.).

A particular area of uncertainty with HSI is the choice of volume for the glass injector chamber. Minimising the volume is desirable to shorten the time for transfer of sample components to the column from the hot, potentially active chamber. However flash-back from vaporised sample overfilling the injector is very undesirable as it leads to broadening of the solvent peak, losses of sample components and contamination of the injector. Despite strong advice that chambers should be about 1 ml for 1–2 μl injections [10, 20], several studies on particular HSI designs have shown very good results with smaller volumes, down to 0.25 ml [13, 14, 17]. As indicated above, the choice of solvent and carrier flow are important factors in reducing vapour volume and back flow. High inlet pressures will also reduce the volume of vapour. Iso-octane vaporised at 250 °C and 100 kPa has a gas volume of 0.13 ml/μl liquid compared to 0.53 ml/μl for methanol. These volumes are approximately halved at 300 kPa. Inserts with a taper at the needle entry can reduce flash back [13, 20]. A 0.53-mm pre-column also provides some buffer volume to partially absorb the pulse of solvent vapour.

The use of deactivated glass wool in the glass liner is sometimes advocated to prevent entry of involatile coextractives into the column. However the additional surface area increases the probability for degradation or adsorption of labile pesticides. A pre-column should be an adequate substitute for glass wool in minimising contamination of the analytical column.

3.3 Programmed Temperature Vaporiser (PTV)

The PTV injector design resembles a compact HSI but PTV operation is similar to the OCI in that injection takes place at low temperatures. However more flexible options are available for the injection and volatilisation conditions [11, 21–23]. The PTV injector zone (5–10 cm) is of low thermal mass which can be rapidly heated and cooled independently of the column oven. There are several configurations and operating modes for the PTV including the following which are suitable for analysis of pesticide residues in water extracts:

a) On column PTV
An insert allows mounting a 0.53 mm i.d. precolumn into the injector so direct injections can be made as for OCI.
b) Cold splitless injection
Injection into a glass insert (1–2 mm i.d.), either empty or with silanised glass wool or other packing (see Fig. 3). Most designs also incorporate a split line at the base of the insert. During sample introduction into the cold injector the split valve is closed. Then the sample is evaporated and transfered into the column. Reopening the splitline after a delay is useful to reduce ghost peaks and other artefacts from thermal decomposition of involatile coextractives.
c) Cold splitless injection with solvent venting
The sample is injected into the cold injector. The glass insert should have baffels or be packed with silanised glass wool. The start temperature of the injector is held below the boiling temperature of the solvent and the solvent is vented through the open split.Then the split valve is closed and the sample is evaporated by flash heating (e.g. 12 °C s^{-1}). After the sample transfer into the

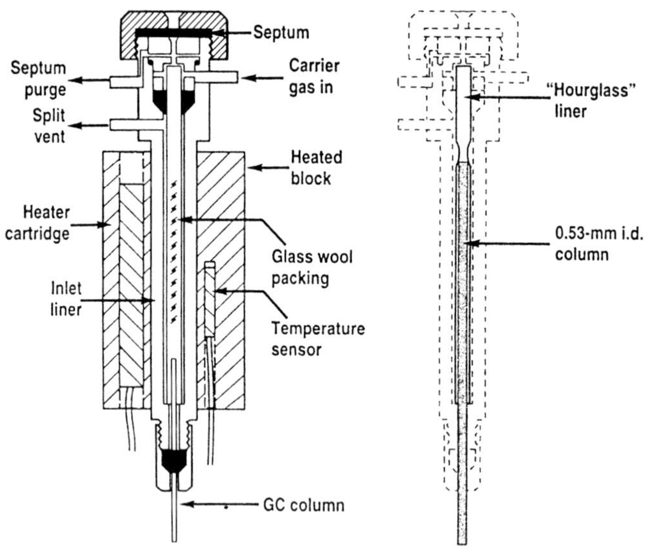

Fig. 3a, b. Programmed temperature vaporiser. **a)** split-splitless mode with glass insert. **b)** On-column injection mode
Reproduced with permission from Ref. [23]

column is finished, the split valve is opened and the injector heated to a second temperature plateau to remove high boiling substances from the injector.

Large volumes of extract can be introduced (25–250 μl) with solvent venting provided the liner type, injection rate and temperature are adjusted to control the evaporation rate of solvent and a retention gap is fitted [24–26].

A main disadvantage of this injection technique is the possible loss of the more volatile components during solvent venting. These possible losses have to be carefully checked during the development of an analytical protocol.

The separate programmable heating of the PTV injector allows for various heating regimes after the injection phase. The most common strategies are:

a) Rapid heating (5–20 °C s^{-1}) to the initial column temperature followed by tracking of the column temperature program. With the on-column PTV configuration this is functionally identical to OCI.
b) Rapid heating (5–20 °C s^{-1}) to the final column temperature followed by a short hold (2–5 min) and then cooling. This may reduce thermal decomposition of involatile coextractives retained in the injector.

As vaporisation of medium- to low-volatility components out of the injector may not be fast enough to achieve the narrow band desired, it is generally advisable to use a low initial column temperature that will focus components on the head of the column.

3.4 Comparative Studies on Injector Performance for Pesticide Residue Analysis

The fundamental superiority of OCI to HSI in reproducibility, recovery of high boiling components and reduced thermal decomposition has been established for a wide range of compounds in various HRGC systems. An early study on phenylurea herbicides established that some were susceptible to thermal degradation with HSI but could be determined using OCI [27]. The PTV has been less rigorously studied but broadly has given results closer to OCI than to HSI. Stan and Muller [28] compared the performance of the 3 injectors in manual and automated modes for HRGC of 23 organophosphorus insecticides (OPs) and a set of *n*-alkanes covering the same elution range (up to C28). Molecular weight discrimination of the HSI relative to OCI was less than 10% up to C28. For stable, low to medium m.w. OP's the responses (relative to internal standard) were similar with all 3 injectors although variability was higher for HSI of the O-Ps, particularly with manual injections. However responses from HSI for particular polar OPs including trichlorfon, dimethoate and dioxathion were much lower and more variable than from OCI. The PTV gave similar results to OCI except for trichlorfon, the most labile pesticide studied. The decomposition product of endrin, endrin aldehyde, was present to about 10% using HSI but was absent with PTV [17]. The HSI was operated at 280 °C to reduce discrimination but late eluters such as coumaphos still gave lower and more variable responses than with PTV. However aminocarb, atrazine and simazine were equally well determined using HSI or PTV. In both the above studies the peak area reproducibility with OCI or PTV was 1–2% RSD versus 2–10% for HSI. Another study comparing the 3 injectors showed major losses of carbamate insecticides including aminocarb when using HSI (220 °C, glass wool packed liner) while no degradation was observed with OCI [29]. The PTV showed some thermal decomposition of the carbamates, particularly if the insert was packed with deactivated glass wool. Evidence was presented indicating sample transfer to the column from an empty PTV insert was virtually complete before temperature ramping, presumably in aerosol form.

Injectors must be judged not only on their performance with standards but also on their ability to deliver reproducible results when analysing large numbers of sample extracts. Coextractives deposited in the injector region can adsorb or degrade sample components and also lead to ghost peaks or other indirect interferences in the chromatograms. On the other hand priming or protective effects of coextractives can lead to improvements to pesticide responses or resolution due to deactivation of active sites in the injector or column. A study of the stability of response for 4 sulfur containing pesticides over 100 injections in a cleaned-up plant extract illustrated some of these effects [30]. OCI gave the best results for the labile pesticides but the fall off in response and resolution was rapid after about 30 injections. This was attributed to the build up of an involatile layer in the pre-column. The HSI and PTV (both with glass wool in the inserts) initially totally degraded the 2 carbamates but were satisfactory for the 2 OPs. The HSI gave some response to 1 carbamate after about 20 injections but performance for all compounds deteriorated rapidly after 40 injections. The PTV also exhibited priming effects but gave stable performance after 15 injections which was retained with only a slight fall-off over the 100 injections.

The effects of coextractives will be strongly dependent on the sample matrix and the precise configuration of the injector. Metal containing pigments, such as chlorophylls, hemes and humic complexes are particularly damaging to injector performance and are likely to be present in many highly concentrated surface water extracts. Plant cuticular ketones probably contribute to priming effects but are not likely to be present at high levels in water extracts. It is not realistic to expect reliable HRGC performance from any injector if large numbers of very dirty extracts are being analysed. However, the trade off between spending time on sample clean-up or on instrument maintenance can only be assessed empirically for a given set of analyses (matrix, pesticides, detection limits).

3.5 Choice of Injector and Operating Mode for Analysis of Water Extracts

The following overall recommendations are based on the considerations outlined above but in many areas remain subjective and will need to be followed in the light of the problem at hand and available equipment.

HSI can be still recommended as a rugged technique suitable for most pesticides. The best performance for labile pesticides will be obtained with a 0.25 ml empty tapered glass insert operating at high linear flow combined with a 0.53 mm i.d. precolumn. The injector temperature should be kept at the minimum to retain adequate responses for late eluters – some discrimination is acceptable provided reproducibility is not affected. A temperature of 220 °C should be sufficient for most pesticides. A high boiling solvent such as isooctane or toluene is advantageous and will be adequately separated from early pesticides. Autoinjectors will provide the best reproducibility and a fast injection mode is recommended. Injection volume should be limited to 1–2 µl.

OCI or On-column PTV is recommended where the analysis has to encompass the widest range of pesticides including those too labile for HSI. Autoinjection (1–2 µl) with a standard syringe into a 1 m by 0.53 mm precolumn is rugged and will provide high reproducibility. Larger volumes can be injected with longer pre-columns and slower injection rates. OCI should give reliable performance on clear drinking water extracts but the pre-column may require frequent replacement with extracts of more complex water samples.

PTV with a narrow bore glass insert has flexibility that allows adjustment of the injection parameters to suit a variety of applications. Injection of 1–5 µl into an empty insert with pre-column will give performance close to OCI but the solvent vent should not be opened. Packed insert PTV is the most tolerant injection technique for dirty samples although some losses of labile pesticides may occur. Venting regimes can be introduced to protect the column and detector from solvent and coextractives.

Large volumes of solvent can be injected using OCI or PTV under carefully controlled conditions which offers an alternative to evaporation of extracts prior to GC with the attendant problems of possible losses of volatile analytes and of volume control.

3.6 Coupled Liquid Chromatography – HRGC (LC-HRGC)

This is the logical extension of injection of large volumes of extract. In this discussion we consider the issues involved in direct coupling of HRGC to automated solvent based technologies such as solid phase extraction (SPE), cartridge chromatographic cleanup (CCC), high performance liquid chromatography (HPLC) or size exclusion chromatography (SEC) that may be used to extract water samples or clean-up extracts. Grob recently has published a monograph on LC-GC [31] and a review of the subject [32]. The advantages of LC-GC include:

1. Full automation of extraction of water samples and HRGC analysis (ground water and drinking water).
2. Full automation of cleanup of sample extracts and HRGC analysis (surface and waste waters).

3. Smaller volumes of water samples requiring collection, transport and analysis. Refering to the example given in the Introduction (Sect. 1), direct processing of water by LC(SPE)-HRGC theoretically would require only a 2–4 ml sample to give equivalent LODs to off-line processing of a 1 litre sample.
4. Reduced sample transfers and evaporations that may lead to losses or contamination.
5. Reduced solvent consumption in the microbore HPLC or SEC systems required for effective on-line LC-HRGC.

The basic technology to achieve LC-GC exists in the form of large volume injections using the OCI or PTV. Some of the requirements have been discussed above for injection of 25–200 μl of non-polar to moderately polar solvents using these injectors. For OCI the key component is the pre-column. These can be of moderate length (2–8 m) provided that the solvent evaporation rate is balanced with the injection rate by adjusting the column temperature and carrier flow (concurrent solvent evaporation, [33]) to avoid overloading of the pre-column with solvent. The solvent also must wet the surface of the pre-column which requires deactivation with phenyldimethyl-silyl (PDMS) rather than TMS groups if solvents more polar than alkanes are to be introduced. The solvent evaporation rate needs to be 50–100 $\mu l\,min^{-1}$ for practical LC-GC coupling. This can be achieved if the temperature of the pre-column is about 20 °C above the boiling point of the solvent but volatile components may be lost through lack of a true solvent effect. Water will not wet deactivated FSOT so reversed phase LC eluants are problematic for concurrent evaporation [34, 35]. Azeotropes of water with *n*-propanol assist evaporation and also wet PDMS deactivated capillaries [35] while use of high boiling co-solvents such as butoxyethanol can regain the solvent effect for good chromatography of more volatile components [36]. Maintaining a high gas flow against the restricting effect of the solvent can also assist evaporation which has been achieved either by flow control rather than pressure regulation of the carrier gas [36] or by introducing a split vent between the precolumn and analytical column [37–39].

A major problem for coupling LC to GC is the rapid loss of deactivation of the precolumn that occurrs when solvents containing moisture are routinely introduced. This leads to peak tailing and losses of polar compounds [12, 39]. Non-aqueous solvents are therefore advantageous for the concurrent solvent evaporation LC-GC coupling technique.

Relatively standard auto-injectors with a flow-through syringes have been adapted as OCI interfaces for automated normal phase HPLC-HRGC [40, 41]. Brinkman and colleagues have developed several systems for automated SPE of water samples direct coupled to HRGC. Organochlorine insecticides at $ng\,l^{-1}$ levels were recovered from water using an ODS-silica cartridge and eluted on-column with hexane [42]. Triazine herbicides [43] and organophosphorus insecticides [39] were also recovered with high reproducibility (RSD 2–6%) at sub $\mu g\,l^{-1}$ levels using SPE cartridges or disks but required ethylacetate for elution. It was found essential in these methods to eliminate all traces of water from the SPE system before elution onto the GC pre-column. This was accomplished by nitrogen gas flushing [39, 43] or in-line drying on a sodium sulfate column [43]. Atrazine in water at ng l^{-1} levels also has been determined using ODS-SPE on-line to HRGC with methanol-water-1-propanol elution and optimised concurrent evaporation [44].

LC-GC clean-up procedures are also of interest for the analysis of surface waters. An automated sample clean-up system using silica cartridges has been coupled to HRGC using OCI of hexane-toluene eluants onto a pre-column with solvent vent [45]. Extracts of organochlorine and synthetic pyrethroid residues in surface waters at ng l^{-1} levels were cleaned-up with high recoveries and reproducibilities. Size exclusion chromatography has been miniaturised and coupled to HRGC using a similar OCI interface [46]. Applications to food analysis were demonstrated but the system also could have considerable application to analysis of surface waters where lipids, pigments and humic substances can cause interfering peaks and rapid deterioration of HRGC performance.

Although the advantages of on-line LC-GC for water analysis appear substantial, further research is required to demonstrate that it can become a routine screening technique. Even where the technology has proved feasible, there are drawbacks including the reduced flexibility, lack of reserve extract for confirmatory analyses and increased probability of faults seriously affecting sample throughput. In many cases parallel operation of automated extraction/clean-up and HRGC systems may be the most effective use of what remain rather complex and idiosyncratic instrumental resources.

4 Selective Detectors for HRGC

The development of electron capture as a selective detector for GC provided the first effective tool for the investigation of the environmental fate of pesticides. Although several other selective detectors were developed in the 1960s and 1970s, there have been few major advances in the basic technology over the past 10–15 years apart from in GC-MS and atomic emission detection (both covered in other chapters in this volume). Earlier reviews are still relevant to the principles of operation of the main selective detectors used for GC of pesticide residues [47, 48]. Advances in the 4 detectors discussed in detail below have principally been incremental improvements in reliability and performance, particularly for use with HRGC. The narrower peaks and generally lower noise levels in HRGC lead to lower detection limits over packed columns [49], although detector volumes and time-constants have to be reduced to avoid peak distortion.

4.1 Detector Characteristics

GC detectors can be classified as having responses dependant on sample mass flow as for flame detectors which consume the sample or on sample concentration as for the electron capture detector. However the use of make-up gas (20–30 ml min^{-1}) with HRGC leads to rather uniform sample concentrations in the gas through any of the detectors. It is therefore convenient to compare detector responses all on a mass flow basis. The basic parameters that characterise the performance of a detector are:

4.1.1 Sensitivity

Sensitivity can be expressed as the response factor – the amount of signal produced per unit mass of sample. It is the slope of the calibration line. For example this can be expressed as the peak area in integrator counts per ng pesticide injected ($\mu\text{V s ng}^{-1}$). Response factors are useful for comparing the responses of different substances on a given detector but are of limited value for comparing different detectors even of a similar type as the measurements incorporate various instrument specific factors such as amplifier gain which influence the signal magnitude.

4.1.2 Minimum Detectability

By also considering the noise component of the signal a more useful definition is obtained which relates directly to the analytical requirements of detectability.

The minimum detectability for a detector using a given compound is defined as the sample flow rate required to give a signal of twice the noise (measured peak(+)–peak(–).

$$\text{Minimum Detectability} \quad MD = 2S_n M_x / S_x W_x \text{ pg s}^{-1} \tag{9}$$

where:
S_n = baseline noise level (mV peak-peak)
M_x = mass of test compound injected (pg)
S_x = peak height response (mV)
W_x = peak half width (s)

Where a detector responds to a particular element the minimum detectability can be expressed in elemental terms e.g. for an FPD as g phosphorus per second.

Similarly the minimum detectable quantity, is defined as the amount of compound that will give a signal twice that of the noise:

$$\text{Minimum Detectable Quantity} \quad MDQ = MD\, W_x \text{ pg} \tag{10}$$

$$= 2S_n M / S_x \text{pg}$$

Column bleed may contribute to the background noise, although detector characterisation is generally carried out under GC conditions where bleed is minimised. Note the MDQ is not the same as the method limit of detection, LOD, which must be determined from the GC responses of untreated control and spiked samples carried through the complete analytical method. Only for very clean extracts and pesticides with high recovery will the LOD for the method approach the MDQ for the GC.

4.1.3 Linearity

The linear range and the related dynamic range are important characteristics of a detector for quantitative work. The linear range is the range of sample flow rates over which the response factor remains approximately constant. Beyond this, saturation and other effects result in a drop off until response factors are drastically reduced at the end of the dynamic range. Log-log plots of response versus amount injected are commonly used for these assessments although such plots are rather insensitive to minor deviations. A linear range can never be assumed particularly when working with GC of pg–ng amounts of labile compounds where injector or column losses also may be a factor. Multi-point non-linear calibrations will frequently be required to accurately represent the response curves for pesticides even in nominally linear regions of selective detector operation.

4.1.4 Selectivity

The selectivity of a detector for two compounds is the ratio of their respective response factors. Where the ratio is small for a wide range of compounds the detector is not selective, such as with flame ionisation. As far as the detection mechanism depends on particular atomic species then selectivity can be expressed in elemental terms as the ratio of the elemental minimum detectabilities e.g. the high selectivity of the flame photometric detector for phosphorus with respect to carbon. In many respects high selectivity is more important than sensitivity for HRGC detection in environmental analyses where coextractives frequently will be limiting factors in the ability to detect or quantitate analytes.

4.1.5 Detector specifications

Table 2 summarises the characteristics of the 4 selective detectors in common use. The specifications are typical values for standard models available on commercial instruments. The detailed pıinciples of operation and characteristics are outlined in the following sections.

Table 2. Performance of selective detectors for high resolution gas chromatography

Detector	Element or functional Group	Minimum detectability ($pg\,s^{-1}$)	Selectivity[a]	Linearity
Electron capture	CF_3, Cl_2, Br	5[b]	10^4	10^4
	Cl_3, Br_2, nitro	0.5[c]	10^5	10^4
	poly-halogen	0.05[d]	10^6	10^4
Flame photometric	P	0.5	10^6	5×10^4
	S[e]	5	10^5	5×10^2
Nitrogen-Phosphorus	N	0.2	10^4	10^5
	P	0.1	10^5	10^4
Electrolytic conductivity	F, Cl, Br	1	10^6	10^6
	N	1	10^6	10^4

a With respect to C/H/O compounds
b 2,4-D methyl ester
c parathion
d lindane
e quadratic response

4.2 Electron Capture Detector (ECD)

The simplicity of operation and extremely high sensitivity to polyhalogenated compounds has kept the ECD as an important selective detector for environmental analyses. Early detectors using tritium β sources and constant voltage operation have been superceded by designs using ^{63}Ni sources which can be heated to 400 °C and pulse modulated (constant current) operation which greatly extends the linear range [50]. The principal changes over the past 15 years have been to reduce the cell volume for use with capillary columns and to improve operation with nitrogen moderator/makeup gas. Figure 4 illustrates a modern concentric cylinder ECD design with low volume cell [51]. A plasma of thermalised electrons is created in the upper cell by β irradiation from the ^{63}Ni plated foil. Pulses of − 50– −70 V are applied to the cell which move the electrons to the lower collector where they are detected by an electrometer circuit. Electron capturing compounds entering the cell reduce the quantity of free electrons collected at each pulse. A constant current is maintained at the collector by use of a feed-back circuit from the electrometer output to the pulsar. The increase in the pulse rate required is proportional to the concentration of compound in the cell (Eq. 11).

$$\text{Pulse rate} \quad f = a_x C_x / K + f_0 \tag{11}$$

where: f_0 = base pulse frequency for carrier gas only (1–5 kHz)
a_x = rate constant for electron capture by the compound
C_x = concentration of compound
K = instrumental constant dependant on activity of source

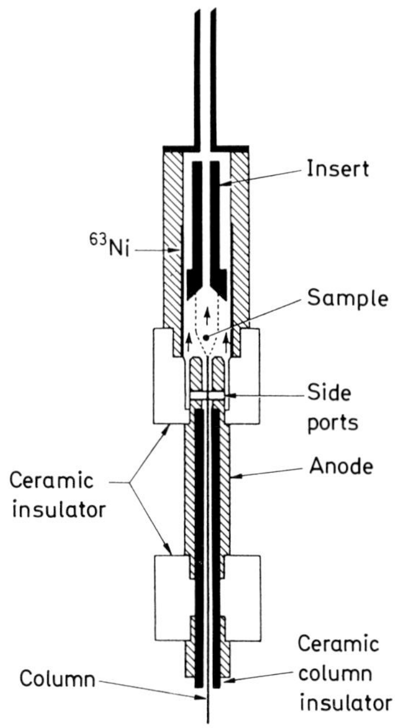

Fig. 4. Low volume electron capture cell Reproduced with permission from Ref. [49]

Pulse modulation, together with optimised cell geometry, can provide a linear range of up to 10^4 versus only 100 for DC mode. Stability of operation is also improved. Linearity is further improved by use of 90% argon + 10% methane rather than nitrogen as moderator gas. The ECD is extremely efficient for compounds with high electron capture rate constants such as organochlorine insecticides which undergo nearly complete electron attachment in the cell, leading to MDQs in the sub-pg range.

The advantage of the very high sensitivity of the ECD to polyhalogenated compounds is off-set by its relatively poor selectivity. Although hydrocarbons, apart from some polynuclear aromatics, give no response, a wide range of organic compounds do respond to a greater or lesser extent. Furthermore the ECD sensitivity is markedly lower for pesticides with lower halogenation levels (see Table 2), further reducing selectivity against coextractives. The detector is of little use for monochloro compounds. The wide range of relative sensitivities of the ECD to different compounds means that chromatograms must be interpreted with care.

4.3 Nitrogen-Phosphorus Detector (NPD)

Various designs of the NPD have alternative names such as the alkali thermionic detector, thermionic specific detector or alkali flame ionisation detector. All have the same operating principle of a hydrogen flame or plasma acting on an alkali salt source such that of nitrogen and phosphorus containing compounds are selectively ionised while the response to carbon is greatly reduced. Early designs based on modified flame ionisation detectors, which used the hydrogen flame to heat the alkali source, have largely been superceded by heated bead designs where separate electrical heating is used to generate a hydrogen rich plasma around the alkali source. These designs have superior minimum detectabilities for nitrogen and are more stable, both important characteristics for automated analyses of extracts for pesticides.

The alkali source is positioned above the jet where carrier + makeup (nitrogen or helium) gases are premixed with a low flow of hydrogen (1–5 ml min^{-1}). In one design the alkali source is fastened to the collector but operation is otherwise similar. A moderate flow of air (100–200 ml min^{-1}) is introduced to the detector chamber. The alkali source is a nichrome or platinum filament coated with a bead of ceramic or glass containing silicates of rubidium, caesium or potassium which is negatively polarised. The heating current through the filament is adjusted so that the hydrogen combusts in a relatively low temperature plasma around the bead thereby ionising alkali atoms which give rise to a low level background current at the electrometer collector. The mechanisms for generation of the N–P response have not been fully elucidated but appear to arise mainly from reductive decomposition of compounds in the plasma followed by surface reactions on the bead generating negative species such as CN^- and PO_2^- [52, 53]. The alkali salt has a role in lowering the work function for the surface ionisation process. The response of the NPD is to a degree tunable by varying the hydrogen flow, polarising voltage and bead temperature. Higher bead temperatures and column flows lead to increased response to P and reduced response to N. An alternative mode has been reported for detection of nitro groups where a special high alkali content bead is used without hydrogen [54].

The NPD is extremely useful for pesticide residue screening due to the low detection limits for N and P and the high selectivity against C/H/O compounds (Table 2). However the response to each element depends on the compound. The N response is higher the more N–C bonds versus N–H, N=C or N–O while phosphorodithioate esters are more responsive than phosphorothioates, phosphates, or phsophoramidates. Calibration is required for each pesticide of interest. NPD chromatograms also need careful evaluation to differentiate between N and P containing compounds because of the low N/P selectivity.

The NPD is not as stable or reproducible a detector as the FID or FPD. The characteristics of the alkali sources change with time leading to reduced sensitivity and higher noise. This process is exacerbated by high column bleed, derivatising reagents, and chlorinated or aromatic solvents. The NPD is not suitable for use with columns containing cyano-silicone stationary phases unless the bleed levels are very low. In some designs P compounds are prone to marked tailing particularly when the detector is tuned for maximum N response. This seems to be due to adsorption and gradual release of P by the bead. The NPD requires relatively frequent calibration of the response to the compounds of interest for quantitative work and the alkali sources will need changing every 1–3 months during regular use.

4.4 Flame Photometric Detector (FPD)

The simple technique of measuring element specific luminescence in a flame is based on some rather complex flame chemistry. The detection of sulfur and phosphorus in the FPD depends on production of excited S_2 and POH respectively in a fuel rich flame with measurement of the chemiluminescence at characteristic wavelengths. As this process is very inefficient a photomultiplier system is required to amplify the weak light signals. In the simplest FPD designs column effluent is mixed with makeup + hydrogen and burnt at a jet in a moderate air flow. An interference filter is used to select the appropriate wavelength depending on the element being studied (usually 526 nm for P and 393 nm for S) and the gas flows are optimised for sensitivity. As the formation of S_2 is a biomolecular process the FPD response to S shows an approximately quadratic response to concentration. The FPD has a very high selectivity against C/H/O/N/Cl compounds.

The FPD is a rather simple and very reliable detector but has some problems for residue analysis due to:

1) The relatively high minimum detectability for S of many designs.
2) The poor selectivity in the P mode against high levels of sulfur compounds.
3) The quenching of chemiluminescence caused by coelution of high concentrations of coextractives. This is an insidious problem as the high selectivity of the FPD against carbon means no peaks may be observed for the coextractive components yet they may still affect quantitation of pesticide peaks eluting nearby.

These problems have been adressed in various ways in more recent designs. The sensitivity for S has been improved by careful attention to burner design including use of sealed chimneys [55, 56]. Quenching effects and selectivity are improved in double burner designs [57, 58]. Sample is combusted in the first flame and chemiluminescence measured in the second flame where S and P containing combustion products are reactivated and CO_2 causes less quenching by perhaps a factor of 100. Unfortunately double burner designs have minimum detectabilities about an order of magnitude higher than the best single burner designs. An optimised single-burner design gave a minimum detectability of 5×10^{-14} gP s^{-1} [59]. Detection limits for S and selectivity against C were improved by use of a rare-earth glass filter in a single burner design with optimised gas flows [60].

The responses of the FPD to P or S containing compounds depend on molecular structure [44, 45]. The responses amongst various P or S compounds become more uniform on double burner designs although universal calibration is still not possible.

The high specificity of the P mode makes the FPD particularly useful for screening water samples for organophosphorus pesticides. The possibility of interference from sulfur compounds must be kept in mind and can be checked by re-injecting the sample diluted 2 fold. P containing peaks should decrease in response 2 fold whereas S peaks will decrease about 4 fold. The S-mode is of more limited utility for screening as there are few pesticides where S is the principal heteroatom and the minimum detectabilities may not be low enough for methods to readilly reach sub-$\mu g\,l^{-1}$ LODs in water.

4.5 Electrolytic Conductivity Detector (ELCD)

The basic principle of the electrolytic conductivity detector is combustion or pyrolysis to produce ionisable gaseous species which are stripped into an electrolyte and detected by the change in electrical conductivity of the liquid. Current commercial versions of the ELCD are based on the improvements made by Hall in the 1970s [61] including miniaturisation, improved gas liquid separation and differential bipolar pulse conductivity measurement with non-aqueous electrolytes. The electrolyte is continuously circulated through the separator/conductivity cell and then through an ion exchange resin for purification and adjustment of the pH. A solvent vent is employed at the column exit to avoid carbonising the reactor or overloading the scrubber. Several modes of operation are available including:

1. Halogen selective mode: This mode is the most straightforward in operation. Reductive pyrolysis of the column effluent mixed with hydrogen in a nickel reactor tube at 800–900 °C produces HX from halogenated compounds. These strong acids are well ionised in an *n*-propanol electrolyte whereas ammonia, carbon dioxide and sulfur products are essentially undissociated and give little response. The ELCD offers higher selectivity than the electron capture detector which in many applications compensates for the higher minimum detectability for polyhalogenated compounds (Table 2). An interlaboratory study established recommendations for obtaining consistent and selective halogen response on the ELCD [62].
2. Nitrogen selective mode: Reductive pyrolysis of nitrogenous compounds in hydrogen to give ammonia is catalysed in a nickel reactor tube. Any acids formed are removed in a small column containing a basic scrubber. Conductivity is determined in a water/alcohol solvent. Optimisation of a recent commercial ELCD design for determination of nitrogenous pesticides has been described [63]. The selectivity of the ELCD in the N mode is higher than that of the NPD, particularly against phosphorus and sulfur compounds, but the minimum detectability for N is about 10 fold higher (Table 2).

3. Sulfur selective mode: Oxidation of sulfur compounds in the pyrolyser with added oxygen gas produces sulfur dioxide which can be detected in the conductivity cell using alcohol-water solvents. Hydrogen halides can be removed by silver metal or aluminosilicate scrubbers. The sulfur mode has not been exploited as much as the halogen or N modes and there is little recent literature on optimisation or performance.

The dead volumes in the pyrolyser and scrubber plus adsorption problems with the polar gases have made it difficult to maintain high resolution with capillary columns. Modern ELCD designs can achieve useful results although most applications have used 0.53 mm i.d. columns [63]. The uses of the ELCD most often reported for water analysis have been for volatile halocarbons and organochlorine pollutants. MDQ's for 15 OC insecticides and 19 PCB congeners were 15–30 pg (RSD 13–34%) for HRGC-ELCD versus 2–10 pg (RSD 2–19%) for HRGC-ECD [64]. Only minor tailing was observed on the ELCD channel with the narrow bore column used. The ELCD was found to be more selective than the ECD for analysis of the OC compounds in a variety of environmental matrices including sea water [64].

Modern ELCDs are generally reliable and sturdy detectors particularly in the halogen mode. However they are rather more complex to operate and maintain than other selective detectors and take longer to set up and stabilize. ELCDs are therefore best suited to routine applications where the lowest minimum detectabilities are not required and where they can be run continuously with regular preventative maintenance of pyrolyser, scrubber and electrolyte systems.

5 Multi-Channel HRGC

A large increase in chromatographic information can be obtained by use of multiple detectors and multiple columns in various combinations. Re-injection of samples onto a separate confirmation column/detector combination represents one off-line approach. Multi-channel HRGC uses equivalent or more powerful on-line techniques to provide the additional information within the one sample run. The more complex flow paths, including splitters, do increase the difficulties of maintaining systems that are leak free and of low surface activity. However suitable fittings and deactivated FSOT are now available. Some approaches which are proving useful in residue analysis include:

5.1 Multiple Columns (Splitting at the Injector)

Additional information can be obtained simultaneously by putting 2 columns into the same injector, either directly with a 2-hole ferrule or with a split at the end of a pre-column. A screening system for food used GC1 with 30 m × 0.25 mm SE54 and 30 m × 0.32 mm DB1701 columns both to ECDs and GC2 with the same non-polar and polar columns to NPD and FPD respectively [65]. The wide bore polar column operated at the same head pressure as the narrow bore non-polar column brought the retention times more into correspondance thereby avoiding excessive analysis time. Similar column combinations have been used with ECD and ELCD (halogen mode) for screening of organochlorine pollutants in environmental samples [64] or with dual ECDs for PCB congener analysis [66].

5.2 Column Effluent Splitting

Two or more detectors can be used simultaneously to provide complementary information on each peak in the chromatograms from a single column by installing a post column splitter [67]. Typical screening and confirmation systems for pesticide residues are GC1 with non-polar column split 1:4 to ECD/NPD and GC2 with polar column split 1:4 to ECD/FPD [68]. The data from this configuration is similar to that obtainable from splitting to 2 columns but is rather easier to interpret as the same components will have identical retention times on both detectors (allowing for a small constant off-set from the split lines).

5.3 Two-Dimensional HRGC

2-D GC has been defined as "Two columns of different selectivity are combined such that eluate fractions can be directly transferred from one column to another" [69]. This powerful technique has been reviewed recently by Stan [70] with illustrations of applications to pesticide residue analysis. The transfer of selected retention regions from the first column to the second (heart cutting) allows a greatly increased resolution of complex or poorly resolved components. The second column is generally held in a separately controlled oven compartment but this is not essential for many applications. The key component for effective 2-D HRGC is the simple but effective pneumatic switching system devised by Deans [71] which redirects column flow with low dead volume and minimal surface activity. Commercial versions of the Deans switch are available. Figure 5 gives the flow schematic for a comprehensive 2-D system incorporating multiple selective detectors for residue screening and confirmation [72]. The Siemens 'live-T' version of the Deans switch is set up so the pressure on restrictor 2 controls whether the flow from column 1 is by-passed to the detectors or is directed to column 2. The system either carries out screening analyses on column 1 (non-polar) with 3 detectors or confirmation and enhanced resolution of selected components by heart cutting to column 2 (polar), again with 3 detectors. 2-D HRGC with heart cutting has also proven particularly powerful in

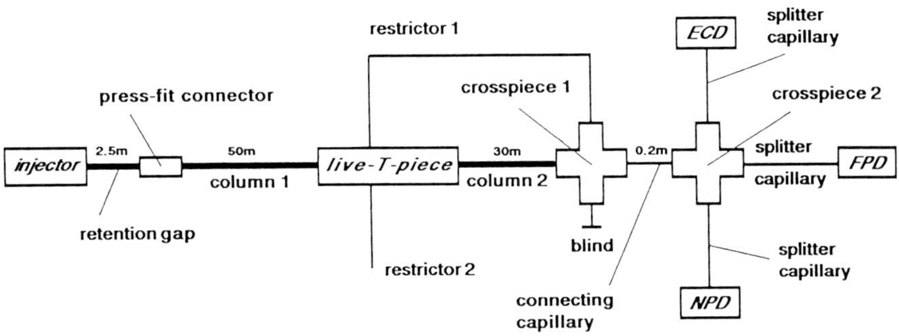

Fig. 5. 2 dimensional HRGC system with multiple detectors for pesticide residue screening.
Reproduced with permission from Ref. [70]

the resolution of the complex mixtures of PCB congeners in commercial formulations or their residues in the environment [73].

Serially coupled polar and non-polar columns can also be used to 'tune' separations [69] by independently adjusting either the temperatures (2 column ovens) or the flows (control of the presure at the junction of the columns). A simple cross-connector system for serial flow tuning has been described [74] which should make the flow tuning technique more accessible for general analysis including screening/confirmation of residues in water.

6 Practical HRGC Operation for Residue Analysis

6.1 Instrument Set-up and Testing

The key requirements are to ensure a leak free, uncontaminated HRGC system which can deliver stable retention times and responses for a wide range of pesticides at pg-ng levels. The ASTM has prepared useful guidelines on the installation of FSOT columns [75] Pre-column and split systems (pre- or post-column) need additional care to ensure the connections are leak-tight, low dead volume, and of low activity. Quartz press-fit connectors may appear the ideal solution but unfortunately the sealing characteristics on various types of FSOT are by no means reliable. Treating the FSOT end with solvent prior to pressing into the connector can help in some cases. However our laboratory has found standard low dead volume stainless steel connectors with graphite\vespel ferrules more reliable. Setting up a post-column split with these connectors requires careful alignment of the column end with the narrow bore FSOT leads but they are trouble free once operating.

Glass injector parts and FSOT pre-columns should be deactivated with diphenyl-tetramethyl-disiliazane (DPTMDS). This can be carried out in the laboratory after thorough cleaning of the parts with detergent and water followed by methanol, acetone and hexane rinses. Fill the FSOT with, or immerse the parts in, a 5% solution of DPTMDS in toluene and stand sealed at room temperature for 60 h. Empty and rinse with toluene followed by hexane.

Effective leak testing of complex HRGC systems requires a thermal conductivity leak detector. Although the expense is considerable, the units are invaluable for finding minute leaks of hydrogen or helium and thus eliminating sources of air which would rapidly degrade the GC performance.

System testing requires injection of mixed standards. The column test mixes used by column manufactures are not convenient for use with selective detectors. A mix of pesticides including both stable and labile compounds of interest should be used on a regular basis to assess injector, column and detector performance. For example a mix containing lindane, acephate, carbaryl, simazine, chlorpyrifos, captan, carbophenothion, iprodione, 4,4′-DDT and azinphos-methyl is suitable for use with ECD, NPD and FPD and contains several compounds that are sensitive to activity in the injector or column. Injections should be made at ng and sub-ng levels. Careful detective work may be required to identify the source of problems. Replacements should be made in the order septum plus injector liner, pre-column then column. Testing should include solvent only and nil injections (temperature programme only) to identify problems with ghost peaks which may require extensive cleaning and bake-out of the injector. Most detector problems can be removed by routine maintenance although ECD's can become subject to persistent high baseline, noise or small linear range which cannot be improved by standard bakeout, with or without hydrogen. Replacement or cleaning of the ^{63}Ni foil may be required but is forbidden in the analytical laboratory in most countries. Manufacturers recommendations and radiation licence conditions should be followed at all times.

Reproducibility of retention times are central to effective screening by HRGC (see below). This will be affected by poor regulation of carrier gas flow or column oven temperature. Aside from actual mechanical or electronic faults, two common causes are septum leaks or inadequate differential between the tank regulator and column head pressures which affects flow control in HSI.

6.2 Calibration

Accurate quantitative work requires multi-level calibration of the HRGC system response to each of the analytes. At least 3 concentration levels are recommended [76] in a nominally linear region from the highest solution concentration expected in sample extracts down to near the detection limit. Non-linear calibrations are frequently required, particularly for the ECD [66]. Thorough calibration is straightforward where a limited range of well resolved components are being determined. However it can become impractical where wide ranging screening is being undertaken for a large number of pesticides, some of which may overlap. The problems that afflict HRGC of labile compounds including injector or column loses and priming effects will influence the accuracy of calibrations. Semi-quantitative protocols therefore may be used for screening with more accurate quantitation of the generally limited range of detected analytes being undertaken during confirmation (see below). Thus single point calibrations at the 0.1–1 ng level may be undertaken daily using standard mixtures divided into sets to cover the full range of pesticides of interest. This will establish the exact retention time table required for initial identifications and allow semi-quantitative estimates of concentrations. Some screening systems mainly rely on more or less historical relative retention/response data tables. This approach puts more reliance on the confirmation systems.

6.3 Internal Standards and Retention Indexes

The use of an internal standard (i.s.) or surrogate standard can increase the accuracy of quantitation in HRGC. Perhaps of greater importance for screening protocols is the quality control they provide from their recovery rate through the analytical system and the relative retention time (RRT) and relative response data to analyte peaks. Suitable surrogate standards should have high recoveries through the sample preparation methods, be highly stable to HRGC, have retention times in mid chromatogram and be separated from the other analytes. It is also useful if they give high responses on all the main selective detectors. Organophosphorus compounds such as malathion or carbophenothion are suitable provided they are very unlikely to be present in the water samples of interest. A second i.s. added prior to evaporation can be useful for volume calculations if samples extracts are being reduced to small volumes (0.5 ml).

In principle multiple i.s. and retention index systems provide even greater accuracy. Bromo-alkanes have been advocated for establishing retention indexes with ECD [77] and a series of alkyl bis(trifluoromethyl) phosphine sulphides have been used for residue screening with several selective detectors [78]. However, with analysis of complex samples such as extracts of surface waters there is increased probability of overlaps with analytes or co-extractives. RRT measurements to a single i.s. have very high reproducibilities on modern instruments making the retention index approach unnecessary for most routine screening.

6.4 Automation

The functions of acquisition, integration and preliminary interpretation of individual chromatograms are well implemented on a range of commercial reporting integrators. However the tasks of calibration for a wide range of components on multiple detectors and the handling of HRGC data from a large number of samples is best handled by a personal computer (PC) based chromatographic data system. A major advantage of these systems is the ability to database the peak information from a sequence of samples and standards. This data can then be transferred to spread sheets or other software packages for more sophisticated analysis and reporting [79]. Such approaches not only systematise data acquisition/analysis and reduce transcription errors but also encourage greater use of control samples such as standards, blanks and spikes. The full power of the HRGC-PC combination can be realised only when auto-injection is also used to provide higher sample throughput and greater reproducibility of injections and retention times.

6.5 Identification Criteria, Confirmation and Data Reporting

When screening samples for a wide range of pesticides, initial identifications will rely on the correspondance of the RRT's to i.s. for the analyte and standard peaks. The full advantage of HRGC can only be obtained from very reproducible retention data. RSD's of less than 0.01% have been reported for RRT's on a variety of commercial instruments. The chances of mis-identifications are greatly increased if RRT's can only be determined to say 0.5%. For polar pesticides some shifts in retention to shorter times may occur with increased amounts injected [80]. Any characteristic peak tailing should match in sample and standard. The peak shapes also need to be carefully studied for evidence of coeluting substances that may be influencing RRT and quantitation. Blank sample chromatograms are required to eliminate laboratory contaminant peaks from consideration. Responses on highly selective detectors such as the ELCD or FPD(P mode) can be taken as good evidence for the presence of particular elements but ECD peaks must be initially regarded with scepticism unless the extracts have been highly cleaned up.

Positive results from the screening matrix of RRT/response on one column/detector combination will require confirmation in all cases except where a known analyte, based on prior confirmed screening is being tracked in a particular set of samples. Some confirmatory data may be obtained in parallel from splitter systems or a separate GC run will need to be carried out. A confirmed match of RRT and response on a more polar column with the same detector will sometimes be adequate. Alternative detection is desirable but not always available for compounds containing only N as a heteroatom e.g. carbaryl and other carbamate insecticides, or containing only halogens e.g. organochlorine insecticides. Confirmation should include improved quantitation by calibrating each sample against standards closely matching (within 50%) the expected concentrations. Ideally there should be a good correspondence of quantitation between screening and confirmation systems ($\pm$ 20%). However this may be affected by limitations in the screening calibration.

Further confirmation of identity may be required, particularly where unusual analytes are detected in samples which have little past knowledge associated with them. GC-MS will generally be the technique of choice [81]. The importance of adequate confirmation was highlighted in some studies on German ground water samples using HRGC screening [82]. In one batch of 81 water samples the HRGC screening chromatograms contained a total of 412 significant peaks of which 237 were tentatively identified as pesticides from RRT data on ECD and NPD. Analysis of the extracts on a second column/detector combination confirmed 132 of the peaks as probable pesticides. However further analyses by HPLC or GC-MS could confirm only 94 pesticide findings (many peaks were equivalent to $< 10\ \mathrm{ng\,l^{-1}}$) of which 37 exceeded $50\ \mathrm{ng\,l^{-1}}$. A set of guidelines for carrying out ground water surveys emphasised the need for confirmatory data in reporting pesticide findings [83].

As well as adequate identification and quantitation of analytes present, the method LODs must be determined for components of interest which were not detected and the status defined of analytes detected but not able to be accurately quantified due to their low levels. Keith has made a good summary of the complex statistical and decision issues relating to setting and reporting of LODs [84, 85]. LOD's are dependent on the whole method and not solely of the HRGC determination step. The key requirement is to obtain data on the background signals from instrument noise, components in the blank and matrix coextractives that limit the ability to measure low levels of each analyte. Rigorous method validation and LOD estimations can be carried out for analysis of pesticide residues in ground water and drinking water. For much more variable matrices such as surface and waste water, LODs may need to be set using more limited data. Conservative approaches to this process have been described [86, 87].

7 Examples of HRGC for Pesticides in Water

A variety of chromatograms are selected to illustrate some particular applications of HRGC to pesticide determination either as separations of standards of particular interest to water analysis or as actual analyses of water extracts.

Figure 6 shows the HRGC-ECD chromatogram of a range of halogenated pesticides in C18-SPE extract of water spiked at $1\ \mathrm{\mu g\,l^{-1}}$ [88]. The good performance of the system is noted for a diverse range of pesticide with widely varying polarities and volatilities.

Figure 7 shows on-line LC-HRGC-ECD chromatograms for direct extract of 1 ml river water samples, unspiked and spiked at $10\ \mathrm{ng\,l^{-1}}$ of organochlorine insecticides [42]. The low detection limits and relatively clean chromatograms are remarkable.

Figure 8 shows a HRGC-NPD chromatogram for standards at a level equivalent to $2.5\ \mathrm{\mu g\,l^{-1}}$ in water from a method to determine triazine herbicides in natural waters [89]. Note the critical separation of atratone, prometone

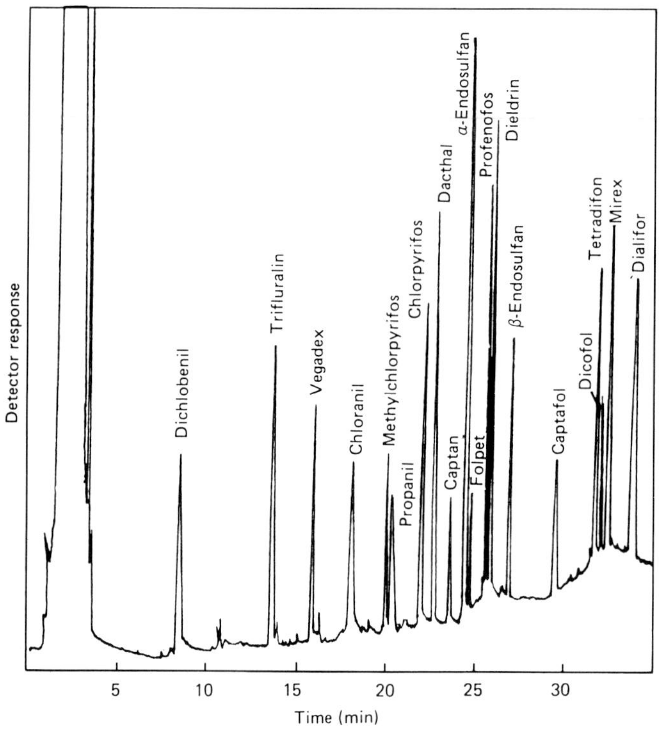

Fig. 6. Halogenated pesticides in C18-SPE water extract by HRGC-ECD. Water spiked at 1 ppb. HSI 260 °C. Column 25 m × 0.22 mm BP5 0.25 µm film, 50 °C 0.8 min, 30 °C/min to 140 °C hold 2 min, 5 °C/min to 280 °C.
Reproduced with permission from Ref. [88]

Fig. 7a, b. On-line LC-HRGC analysis of river water **a** unspiked, **b** spiked with organochlorine insecticides, 10 ppt each.
1 ml of water preconcentrated on C18 pre column, dried with nitrogen and eluted with 50 µl hexane direct on-column to HRGC. Retention gap 5 m × 0.5 mm. Column 25 m × 0.22 mm CP Sil 5CB 0.13 µm film. 90 °C 15 min, 10 °C/min to 120 °C hold 2 min, 10 °C/min to 190 °C, 3 °C/minto 230 °C.
Peak identity: 1, α-LCH; 2, HCB; 3, γ-HCH; 4, heptachlor expoxide; 5, aldrin; 6, heptachlor; 7, 2, 4′-DDE; 8, endosulfan; 9, 4, 4′-DDE; 10, dieldrin; 11, 2, 4′-DDD; 12, endrin; 13, 4, 4′-DDD; 14, 2, 4′-DDT, 15, 4, 4′-DDT.
Reproduced with permission from Ref. [42]

Fig. 8. Triazine standards (1 ng each) by HRGC-NPD. HSI 250 °C. Column 30 m × 0.25 mm DB-1 0.25 mm film, 70 °C 0.5 min, 25 °C/min to 195 °C hold 8 min, 10 °C/min to 200 °C.
Peak identity; 1, prometone; 2, atratone; 3, propazine; 4, simetone; 5, atrazine; 6, prometryne; 7, simazine; 8, ametryne; 9, cyprazine; 10, simetryne; 11, cyanazine.
Reproduced with permission from Ref. [89]

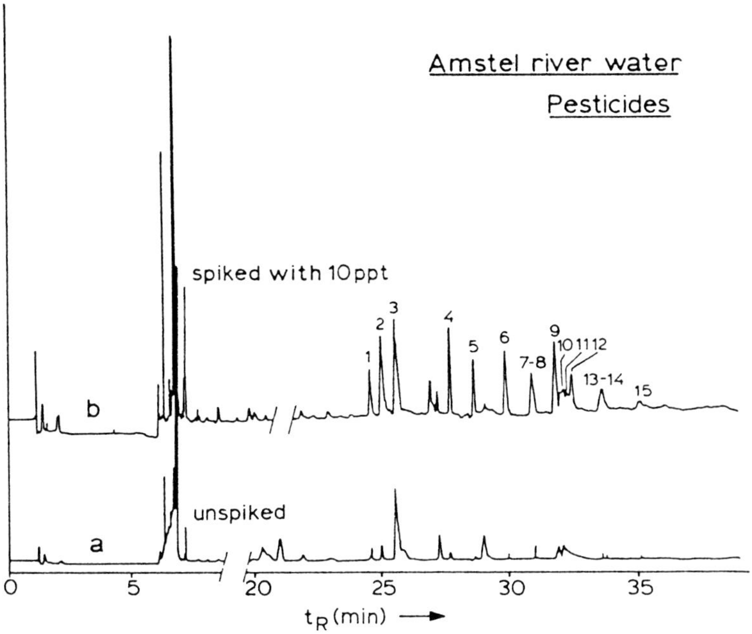
Amstel river water
Pesticides
spiked with 10ppt
b
unspiked
a
1
2
3
4
5
6
7-8
9
10 11 12
13-14
15
0
5
20
25
30
35
t_R (min)

Fig. 7a, b.

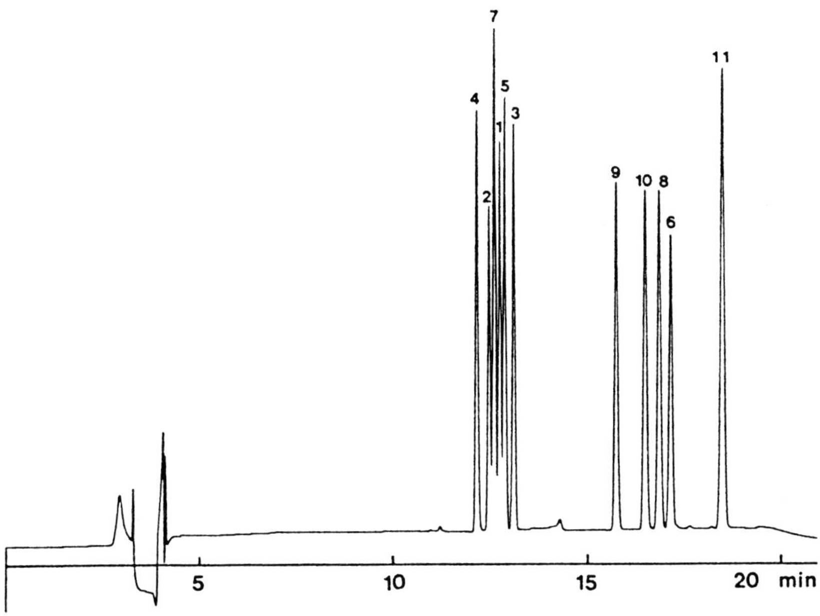
1
2
3
4
5
6
7
8
9
10
11
5
10
15
20 min

Fig. 8.

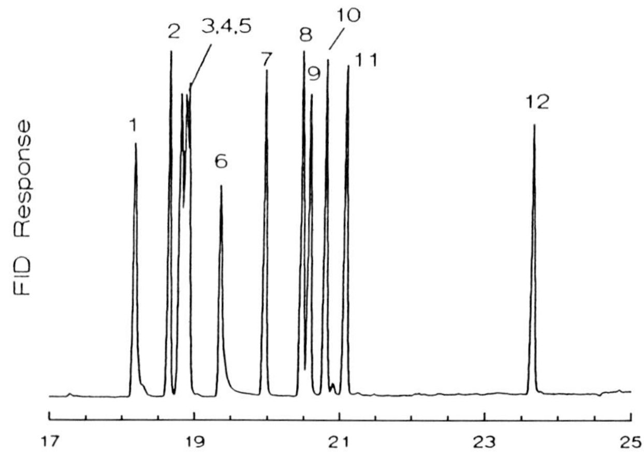

a

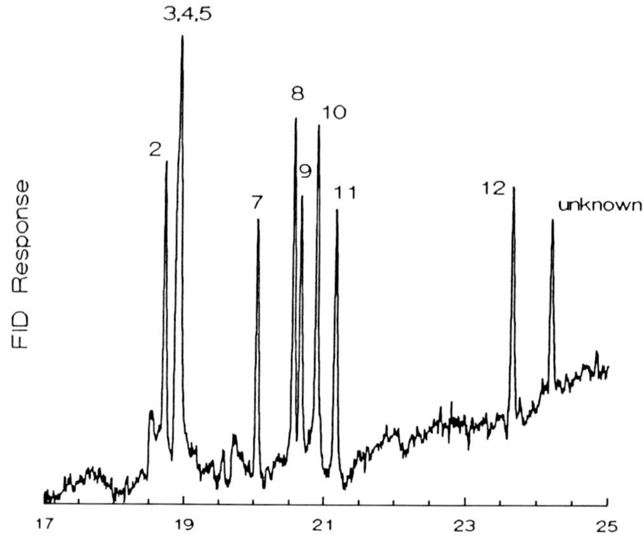

b

Fig. 9. On-line LC-HRGC analysis of aqueous standard containing 12 triazines at (**a**) 0.2 ppb (**b**) 27 ppt. 10 ml of water preconcentrated on polymer precolumn and eluted with 50 μl ethyl acetate through a sodium sulphate drying column direct on-column to HRGC. Retention gap 7.5 m × 0.32 mm. Column 15 m × 0.25 mm DB1701 0.25 μm film 84 °C ca 8 min, 8 °C/min to 285 °C.
Peak identity; 1, atratone; 2, trietazine; 3, simazine; 4, atrazine; 5, propazine; 6, sec-bumeton; 7, sebutylazine; 8, prometryn; 9, simetryn; 10, terbutryn; 11, dipropetryn; 12, cyanazine.
Reproduced with permission from Ref. [43]

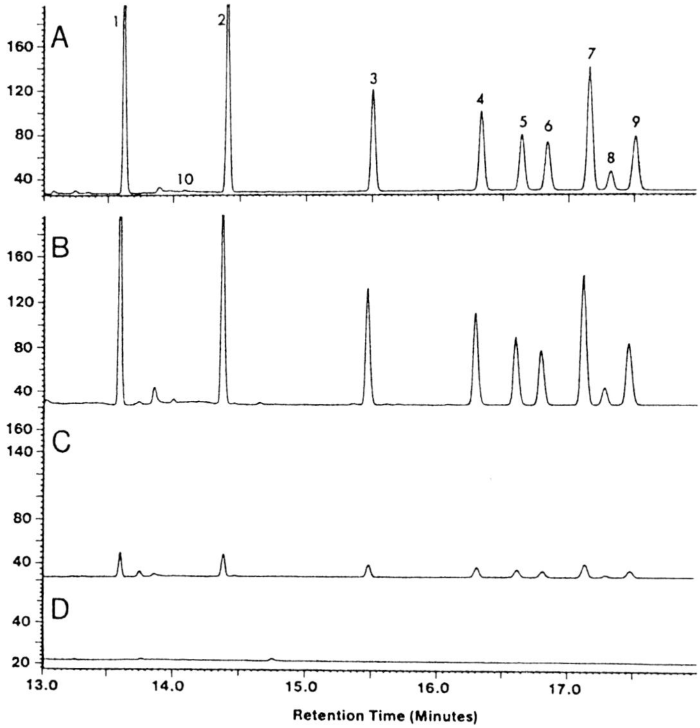

Fig. 10. HRGC-NPD chromatograms for a range of triazole and pyrimidine pesticides. HSI 250 °C. Column 25 m × 0.22 mm SE52/54 0.5 μm film 50 °C 2 min, 20 °C/min to 260 °C.
A. Standard 1 ng each
B, C. Fortified water at 1 and 0.1 ppb levels respectively.
D. Control water
Peak identity; 1, pirimicarb; 2, pirimiphos-methyl; 3, ketone analogue of paclobutrazol; 4, paclobutrazol; 5, flutriafol; 6, hexaconazole; 7, bupimarate; 8, fluazifop-P-butyl; 9, diclobutrazol; 10, ethirimol.
Reproduced with permission from Ref. [90]

(metabolites), simazine and atrazine which required much more polar packed columns.

Figure 9 shows the direct LC-HRGC analysis of a similar mixture of triazines at 0.2 $\mu g\,l^{-1}$ and 27 $ng\,l^{-1}$ from preconcentration of 10 ml of spiked water [43]. This analysis is noteworthy for the low detection limits achieved, even though the non-selective FID was used. Some loss of resolution was found for lower levels of simazine, atrazine and propazine.

Figure 10 shows HRGC-NPD chromatograms for low level determination of a range of triazole and pyrimidine pesticides as standards and in C18 SPE extracts of water [90].

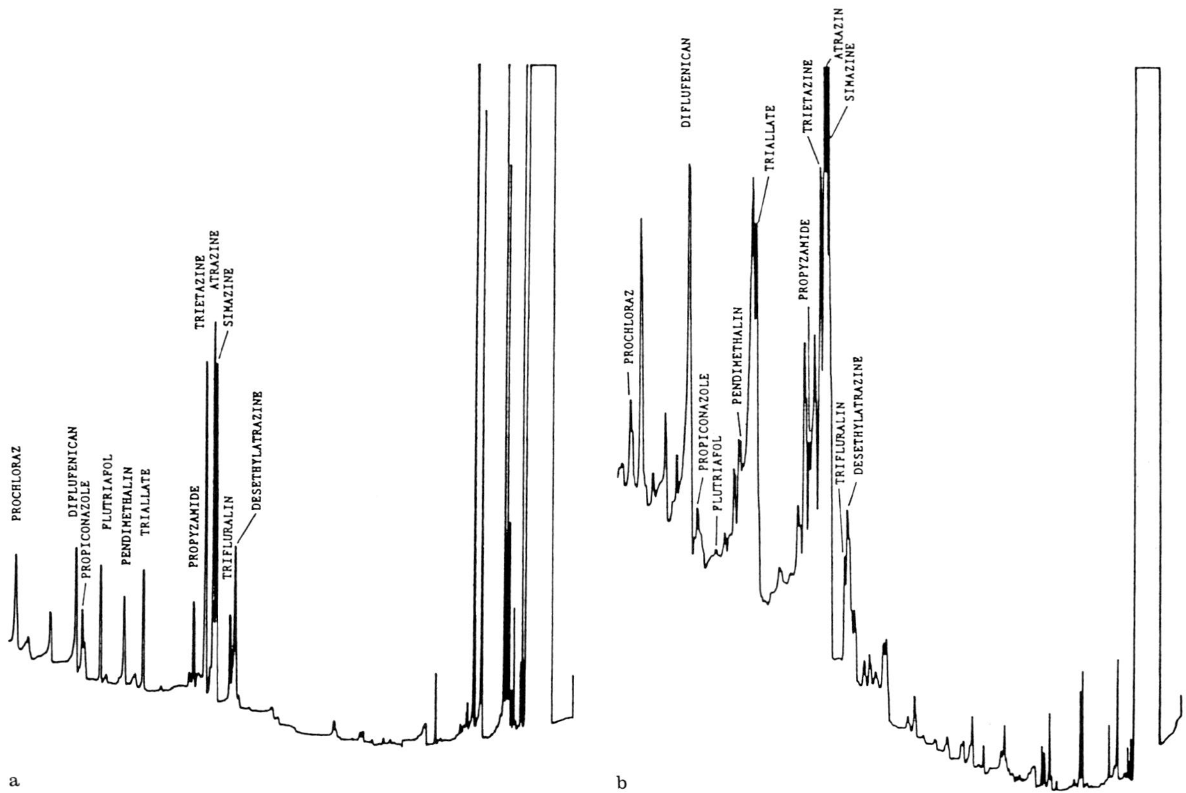

Fig. 11. HRGC-NPD chromatograms for organonitrogen pesticides in dichloromethane extracts of water spiked at 0.1 ppb **a** bore hole drinking water **b** river water. OCI, 30 m DB5 60 °C 3 min, 8 °C/min to 300 °C. Reproduced with permission from Ref. [91]

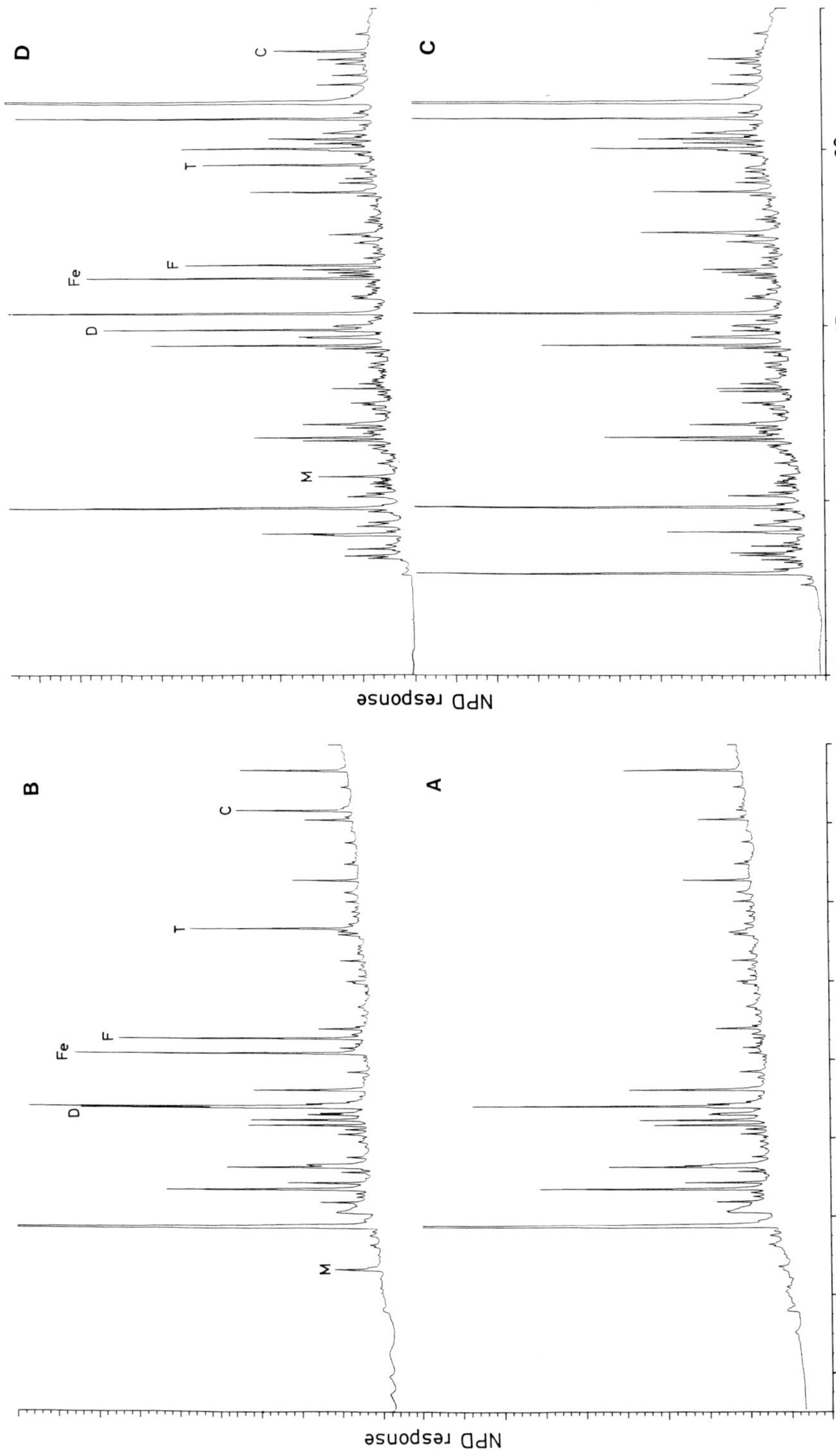
D
C
B
A
M
D
Fe
F
T
C
NPD response
Time (min)
5
10
15
20
6
8
10
12
14
16
18
20
22

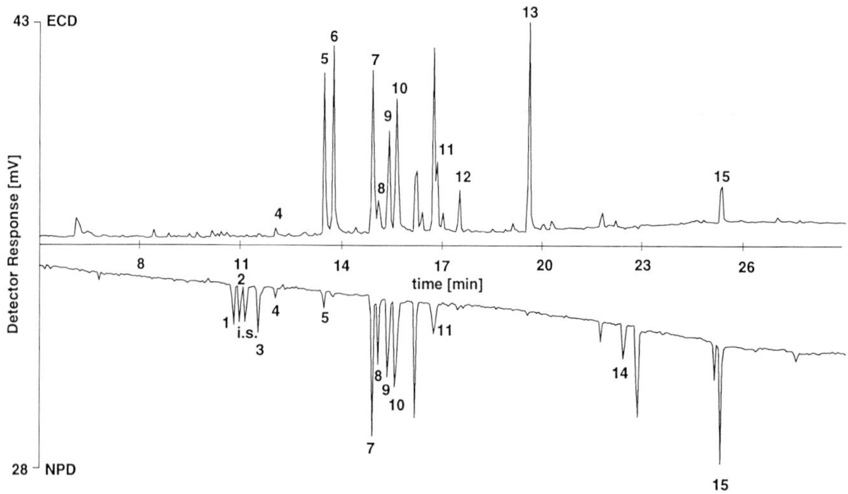

Fig. 13. Neutral pesticides in C18-SPE (disk) extract on ECD and NPD. Water spiked at 0.5 ppb.
HSI 220 °C. Precolumn 1 m × 0.53 mm. Column 25 m × 0.20 mm HP-5, 0.33 μm film, 85 °C 1 min, 40 °C/min to 150 °C hold 2 min, 5 °C/min to 250 °C hold 20 min. Post column split (ca 1:4) to ECD and NPD.
Peak identity: 1, simazine; 2, atrazine; i.s., propazine; 3, terbuthylazine; 4, diazinon; 5, metribuzin; 6, vinclozolin; 7, bromacil + pirimiphos-methyl; 8, phorate sulfoxide; 9, phorate sulphone; 10, cyanazine + chlorpyrifos; 11, pendimethalin; 12, procymidone; 13, oxyfluorfen; 14, hexazinone; 15, azinphos-methyl.

The problems of screening more complex surface waters is illustrated in Fig. 11 where HRGC-NPD chromagrams for a river water spiked at 0.1 $\mu g\,l^{-1}$ with various N-containing pesticides shows a number of serious interferences as compared to a borehole water sample [91].

Figure 12 also shows the range of interferences to be expected from polluted river waters. In this case on-line LC-HRGC was used to establish detection limits for several organophosphorus pesticides in water from the upper and lower Rhine [39].

Figure 13 shows the HRGC screening of a C18 SPE (disk) extract of river water spiked at 0.5 $\mu g\,l^{-1}$ with a range of pesticides using ECD and NPD (this

Fig. 12a–d. On line LC-HRGC analysis of river water using NPD. **a** Rhine (Uster) water; **b** Rhine (Uster) water spiked with organophosphorus insecticides at 0.12–0.19 ppb; **c** Rhine (Lobith) water; **d** Rhine (Lobith) water spiked with OPs at 0.12–0.19 ppb. 2.5 ml of water preconcentrated on SPE disks and eluted (after-drying) with 70 μl ethyl acetate direct on-column to HRGC. Retention gap 5 m × 0.32 mm with solvent vent T. Column 25 m × 0.32 DB5 80 °C 5 min, 10 °C/min to 280 °C.
Peak identity: M. mevinphos; D, diazinon, Fe, fenitrothion; F, fenthion; T, triazaphos; C, coumaphos.
Reproduced with permission from Ref. [39]

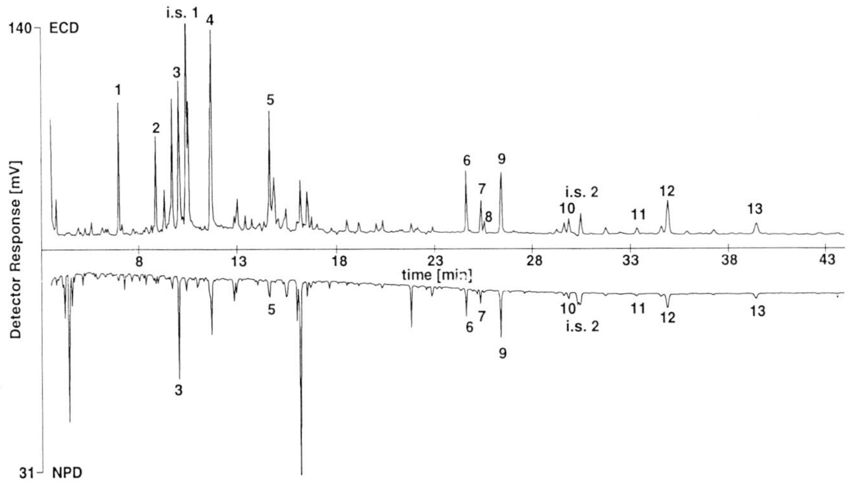

Fig. 14. Acidic pesticides in methylated extract (C18-SPE disk, pH2) on ECD and NPD. Water spiked with benzoic/phenoxy acids (0.5 ppb) and sulfonylureas (5 ppb). GC conditions as for Fig. 13.
Peak identity: 1, dicamba; 2, 2,4-D; 3, triclopyr; i.s.1, 2,4,6-T; 4, 2,4,5-T; 5, picloram; 6, primisulfuron; 7,8, primisulfuron OCH_3 derivatives; 9, chlorosulfuron; 10, metsulfuron methyl; i.s.2, thifensulfuron; 11, chlorimuron; 12, triasulfuron; 13, cinosulfuron.

laboratory). The unnumbered peaks are matrix derived. The advantage of dual detection for confirmation are obvious both where response are strong on each channels e.g. metribuzin or azinphosmethyl and where response on one channel is relatively low e.g. triazines on ECD and vinclozolin or procymidone on NPD.

Figure 14 shows the dual ECD/NPD chromatograms for the methylated extract (C18-SPE disk, pH2) of a water sample spiked with a range of benzoic/phenoxy-acid herbicides (0.5 $\mu g\,l^{-1}$ each) and sulfonylurea herbicides (5 $\mu g\,l^{-1}$ each). The data was developed in this laboratory using a new method of preparing the thermo-stable *N,N'*-dimethyl derivatives of sulfonylureas [92] and analysing them in water or soil [93, 94]. Primisulfuron produces 3 peaks due to some formation of the methoxy isomers (peaks 7 and 8). The HRGC of these derivatives can achieve method LODs of lower than 10 $ng\,l^{-1}$ and 100 $ng\,l^{-1}$ respectively for these 2 classes of herbicides in drinking water and allows for the first time GC-MS confirmation of sulfonylureas as the intact herbicides [93, 94].

8 Conclusions

This review has attempted to provide some basic information on fundamentals of HRGC important to residue analysis, to examine some recent technical advances and to give examples of current achievements in the field of multi residue screening of water. HRGC has become central to detection of low level residue in water although alternative techniques such as HPLC also have important applications. Moore [91] has pointed out that most current methods do not satisfy the ideal performance characteristics required by regulatory authorities for analytical methods to monitor drinking water supplies. Therefore there will be continued efforts to develop multi-residue methods with increased scope, lower detection limits, improved accuracy and precision, higher throughput and lower costs. HRGC will have a place in these developments because of the high confidence in identification and speciation that can be obtained and the scope for further automation such as direct coupled LC-GC.

Acknowledgments. Peter Klaffenbach and Colin Malcolm provided skilled technical assistance in developing HRGC methodology.

9 References

1. Silvis PH (1989) LC-GC 7: 562
2. Kaiser R (1961) Z. Anal. Chem. 189: 1
3. Golay MJE (1958) Theory and practice of gas liquid partition chromatography with coated capillaries. In: Coates VJ, Noebels HJ, Fagerson IS (eds) Gas chromatography. Academic Press, New York, p1
4. Gaspar G (1992) J. High Res. Chromatogr. 15: 295
5. Villalobos R, Annino R (1989) J. High Res. Chromatogr. 12: 149
6. Stark TJ, Larson PA, Dandeneau RD (1985) J. Chromatogr. 279: 31
7. Hyver KJ, Philips RJ (1987) J. Chromatogr. 399: 33
8. Sandra P (1987) LC-GC 5: 236
9. Klee MS, Chang I (1991) J. High Res. Chromatogr. 14: 18
10. Grob K (1986) Classical Split and Splitless Injection in Capillary Gas Chromatography. Huethig, Heidelberg
11. Grob K (1991) On-column Injection in Capillary Gas Chromatography, 2nd edn. Huethig, New York
12. McCabe T, Hiller JF, Morabito PL (1989) J. High Res. Chromatogr. 12: 517
13. Yang FJ, Brown AC, Cram SP (1978) J. Chromatogr. 158: 91
14. Bayona JM, Aparicio X, Albaiges J (1986) J. High Res. Chromatogr. 9: 59
15. Snell RP, Danielson JW, Oxborrow GS (1987) J. Chromatogr. Sci. 25: 225
16. DeVeaux RD, Szelewski M (1989) J. Chromatogr. Sci. 27: 513
17. Penton ZE (1991) J. Assoc. Offic. Anal. Chem. 74: 872
18. Wylie PL, Klein KJ, Thompson MQ, Hermann BW (1992) J. High Res. Chromatogr. 15: 763
19. Lee H-B, Szawiola R, Chau ASY (1987) J. Assoc. Offic. Anal. Chem. 70: 929
20. Grob K, Biedermann M (1989) J. High Res. Chromatogr. 12: 89
21. Schomburg G, Husmann H, Behlau H, Schultz F (1983) J. Chromatogr. 279: 251
22. Vogt W, Jacob K, Ohnesorge PB, Obwexer W (1979) J. Chromatogr. 186: 197
23. Hinshaw JV (1992) LC-GC 10: 748
24. Schomburg G, Husmann H, Schultz F (1983) J. Chromatogr. 279: 259
25. Stan H-J, Christall B (1988) Dtsch. Lebensm.-Rundsch. 84: 375

26. Staniewski J, Rijks JA (1992) J. Chromatogr. 623: 105
27. Grob K (1981) J. Chromatogr. 208: 217
28. Stan H-J, Muller H-M (1988) J. High Res. Chromatogr. 11: 140
29. Müller H-M, Stan H-J (1990) J. High Res. Chromatogr. 13: 759
30. Müller H-M, Stan H-J (1990) J. High Res. Chromatogr. 13: 697
31. Grob K (1991) On-line coupled LC-GC. Huethig, New York
32. Grob K (1992) J. Chromatogr. 626: 25
33. Munari F, Trisciani A, Mapelli G, Trestianu S, Grob K, Colin JM (1985) J. High Res. Chromatogr. 8: 601
34. Grob K, Li Z (1989) J. Chromatogr. 479: 381
35. Grob K, Li Z (1989) J. Chromatogr. 473: 391
36. Grob K, Müller E (1989) J. Chromatogr. 473: 411
37. Vreuls JJ, Cuppen WJGM, Dolecka E, Maris FA, de Jong GJ, Brinkman UA Th (1989) J. High Res. Chromatogr. 12: 807
38. Bicchi C, D'Amato A, Semeraro I, Galli A, Gali M (1992) J. High Res. Chromatogr. 15: 155
39. Kwakman PJM, Vreuls JJ, Brinkman UA TH, Ghijsen RT (1992) Chromatographia 34: 41
40. Ramsteiner KA (1987) J. Chromatogr. 393: 123
41. Fowlis IA (1990) J. High Res. Chromatogr. 13: 213
42. Noroozian E, Maris FA, Nielsen MWF, Frei RW, de Jong GJ, Brinkman UA Th (1987) J. High Res. Chromatogr. 10: 17
43. Vreuls JJ, Ghijsen, RT, de Jong GJ, Brinkman UA Th (1992) J. Chromatogr. 625: 237
44. Grob K, Li Z (1989) J. Chromatogr. 473: 423
45. van der Hoff GR, Gort SM, Baumann RA, van Zoonen P, Brinkman UA Th (1991) J. High Res. Chromatogr. 14: 465
46. Grob K, Kalin I (1991) J. Agric. Food Chem. 39: 1950
47. Holland PT, Greenhalgh R (1980) In: Anson Moye H (ed) Analysis of pesticide residues. Wiley, New York, p 51
48. Dressler M (1986) Selective gas chromatographic detectors. Elsevier, Amsterdam (Journal of chromatography library, vol 36)
49. Wells G (1983) J. Chromatogr. 270: 135
50. Maggs RJ, Joynes PL, Davies AJ, Lovelock JE (1971) Anal. Chem. 43: 1966
51. Wells G (1983) J. High Res. Chromatogr. 6: 651
52. Kolb B, Auer M, Pospisil P (1977) J. Chromatogr. Sci. 15: 53
53. Patterson PL (1978) J. Chromatogr. 167: 381
54. Patterson PL, Gatten RA, Ontiveros C (1982) J. Chromatogr. Sci. 20: 97
55. Greenhalgh R, Wilson M (1978) J. Chromatogr. 128: 157
56. Liu G, Luo X, Zhang S (1988) J. Chromatogr. 435: 327
57. Patterson PL, Howe RL, Abu-Shumays A (1978) Anal. Chem. 50: 339
58. Vogt CR, Kapila S (1978) J. Chromatogr. Sci. 17: 546
59. Verga GR (1992) J. High Res. Chromatogr. 15: 235
60. Driscoll JN, Berger AW (1989) J. Chromatogr. 468: 303
61. Hall RC (1974) J. Chromatogr. Sci. 12: 152
62. Carson KJ (1983) J. Assoc. Offic. Anal. Chem. 66: 1335
63. Fehringer NV, Gilvydis DM, Walters SM, Poole CF (1992) J. High Res. Chromatogr. 15: 124
64. Durell GS, Sauer TC (1990) Anal. Chem. 62: 1867
65. Andersson A, Ohlin B (1986) Var Foeda, Suppl. 2: 79
66. Storr-Hansen E (1991) J. Chromatogr. 558: 375
67. Stan H-J, Goebel H (1983) J. Chromatogr. 268: 55
68. Holland PT, Malcom CP (1992) In: Cairns T, Sherma J (eds) Emerging strategies for pesticide analysis. CRC Press, Boca Raton, FL, p 71

69. Bertsch W (1990) In: Cortes HJ (ed) Multidimensional chromatography. Marcel Dekker, New York, p 74
70. Stan H-J (1992) In: Cairns T, Sherma J (eds) Emerging strategies for pesticide analysis. CRC Press, Boca Raton, FL, p 175
71. Deans DR (1968) Chromatographia 1: 18
72. Stan H-J, Heil S (1991) Fresenius J. Anal. Chem. 339: 34
73. Gunther FR, Chesler SN, Rebbert RE (1989) J. High Res. Chromatogr. 12: 812
74. Toth T, Garay F, Lovasz C (1990) J. High Res. Chromatogr. 13: 840
75. Silvis PH (1992) LC-GC 10: 368
76. ACS Committee on environmental improvement (1983) Anal. Chem. 55: 2210
77. Pacholec F, Poole CF (1980) Anal. Chem. 54: 1019
78. Manninen A, Kuitunen M-L, Julin L (1987) J. Chromatogr. 394: 465
79. Stan H-J (1989) In: Sherma J (ed) Analytical methods for pesticides and plant growth regulators. vol 17. Academic Press, San Diego p 167
80. Ripley BD, Braun HE (1983) J. Assoc. Offic. Anal. Chem. 66: 1084
81. Holland PT (1990) Pure & Applied Chem. 62: 317
82. Pestemer W (1991) Unpublished report. BBA, Braunschweig, FRG
83. Brauch H-K, Schullerer S (1991) Fresenius J. Anal. Chem. 339: 420
84. Keith LH (1991) Chemtech June: 352
85. Keith LH (1991) Chemtech August: 486
86. Conacher HBS Baugh T, Boyer KW, Clower M, Elkins ER, Gross T, Jensen TL, McWeeny DJ, Myreal G, Stephany RW, Woodward B (1990) J. Assoc. Off. Anal. Chem. 73: 332
87. Parker GA (1991) J. Assoc. Offic. Anal. Chem. 74: 868
88. Manes J, Pico Y, Molto JC, Font G. (1990) J. High Res. Chromatogr. 13: 843
89. Lee H-B, Stokker YD (1986) J. Assoc. Offic. Anal. Chem. 69: 568
90. Bolygo E, Atreya NC (1991) Fresenius J. Anal. Chem. 339: 423
91. Moore KM (1991) In: Creaser C, Purchase R (eds) Food contaminants – Sources and surveillance. Royal Society of Chemistry, London, p 177
92. Klaffenbach P, Holland PT, Lauren DR (1993) J. Agric. Food Chem. 47: 388
93. Klaffenbach P, Holland PT (1993) J. Agric. Food Chem. 41: 396
94. Klaffenbach P, Holland PT (1993) Biological Mass Spectrom. 22: 565

GC Analysis Based on Mass Spectrometry

Hans-Jürgen Stan

Institute of Food Chemistry, Technical University Berlin,
Gustav-Meyer-Allee 25, D-13355 Berlin, FRG

A short introduction is given on the theory of mass spectrometry with respect to its outstanding importance for pesticide residue analysis in environmental samples. The origin of molecular ions and fragments from various ionization techniques is discussed with respect to their suitability for interpreting mass spectra as representative for a definite molecular structure. By this means, it is possible to identify an individual pesticide unequivocally. Special applications of highly sophisticated instrumentation, such as high resolution or tandem mass spectrometry are included.

Chemistry of Plant Protection Vol. 11

1 Introduction and Scope

Mass Spectrometry has gained a position of outstanding importance in many areas of organic chemical analysis. The technique can be combined on-line with the most important chromatographic separation techniques applied in trace analysis of biological and environmental samples: capillary gas chromatography and high performance liquid chromatography. Both these "hyphenated techniques" GC-MS and LC-MS are nowadays indispensable for sensitive positive structural identification of pesticides and other pollutants in our environment. These techniques are extremely valuable for the development of analytical methods to meet the low maximum tolerance values as set by the European Community for pesticides in drinking water. Therefore, a brief review of the technique of mass spectrometry is included to survey the basics and to emphasize what makes this technique the unrivaled detection method in pesticide analysis. To date virtually no analytical result can be considered as reliable if it does not include a mass spectrometric confirmation!

2 The Mass Spectrum

The mass spectrum is a plot of the intensity as a function of the mass-to-charge ratio (Fig. 1). The peak with the highest intensity in the spectrum is called the base peak. Generally, the spectrum is normalized to the intensity of the base peak, resulting in relative intensities.

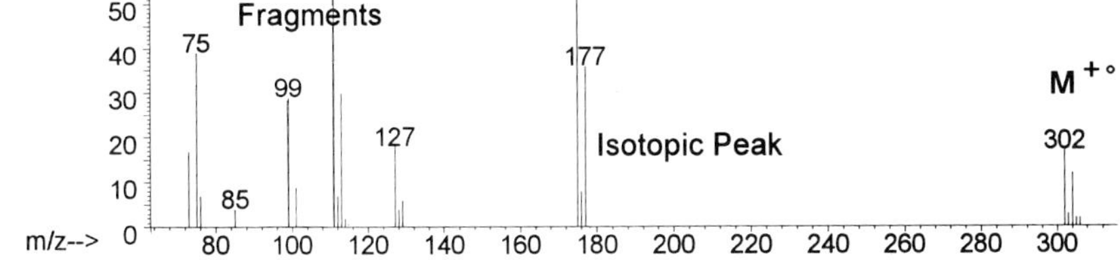

Fig. 1. The mass spectrum

When a molecule is ionized, a molecular ion $M^{+\cdot}$ is produced and this may contain sufficient internal energy to fragment by ejection of a neutral particle N with the formation of a fragment ion $A^{+\cdot}$ or A^+. The original analyte molecule gives a radical-cation as the molecular ion, and the fragment ion may be a cation or a radical-ion. The ejected neutral particle N may be a radical or a neutral molecule

$$M + e \rightarrow M^{+\cdot} \rightarrow A^+ + N^\cdot$$

or

$$M + e \rightarrow M^{+\cdot} \rightarrow A^{+\cdot} + N$$

If the fragment ion has sufficient internal energy, then further fragmentation may occur with the formation of a whole series of fragment ions

$$M^{+\cdot} - N^\cdot \rightarrow A^+ - N_A \rightarrow B^+ - N_B \rightarrow C^+ \cdots$$

or

$$M^{+\cdot} - N \rightarrow Z^{+\cdot} - N_Z \rightarrow Y^{+\cdot} - N_Y^\cdot \rightarrow X^+ \cdots$$

Such a series of decompositions when elucidated from a mass spectrum is called a fragmentation pathway. As shown, the molecular ion $M^{+\cdot}$ may decompose by more than one pathway. The various fragmentation pathways together compose a fragmentation pattern characteristic of the compound under investigation.

The extent to which fragmentation takes place along the individual pathways is determined by the amount of internal energy originally given to the molecular ion $M^{+\cdot}$ and its structure. Hence, the mass spectrum is not simply

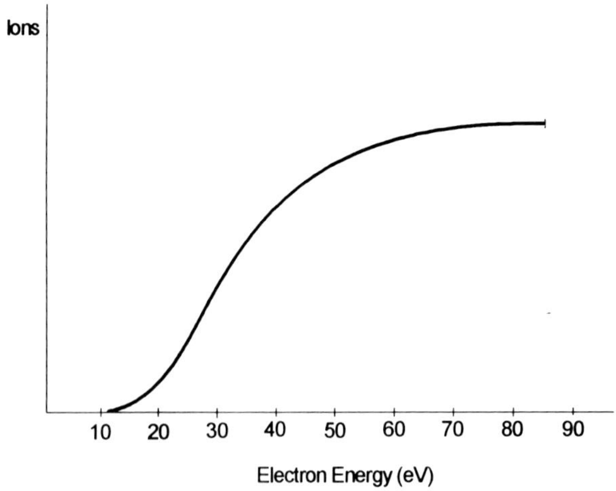

Fig. 2. Total ionization current as function of ionization energy

a fragmentation pattern but its appearance depends upon the energy of the ionizing electrons and also upon the temperature at which ionization occurs.

Ion formation as a function of the energy of ionizing electrons in the electron impact process is illustrated in Fig. 2. The electrons are provided by a heated filament in the evacuated ion source and accelerated through a potential and directed across the chamber where they may hit the analyte molecules. The potential is continuously variable between zero and over 100 eV. Molecular ions begin to appear around 10 eV (ionization energy or appearance potential) but at low intensities. Between 10 and 15 eV fragment ions also begin to appear. Both types of ion increase in abundance following individual abundance curves. Standard mass spectra are obtained at 70 eV because maximum ion yield (total ion current) is observed at this value and mass spectra were found to be reproducible and characteristic of the molecule ionized almost independant of the type and make of instrument.

3 Emergence of the Mass Spectrum

A short overview of the basic aspects of unimolecular ion decomposition should be helpful in understanding the capabilities and limitations of mass spectrometry with electron impact ionization for structure elucidation. This should at the same time convince the analyst that a mass spectrum generally reflects also small changes in the molecular structure and is therefore indicative of this particular chemical compound (with the well-known limitation for structural isomers). Ionization of analyte molecules with 70 eV electrons produces molecular ions whose internal energy values E cover a broad range averaging a few eV. This distribution is described by the probability function P (E) and shown in the upper part of Fig. 3.

The initially formed $M^{+\cdot}$ ions show a wide range of decomposition behavior depending on the internal energy. The probability of any decomposition reaction is expressed as the rate constant k which changes with the internal energy E of the molecular ion. A certain reaction critical energy is necessary to form the various decomposition products that can be observed in the mass spectrum if a rate constant of 10^6 s^{-1} or greater is reached. This is due to the fact that ion-source decomposition can only be observed for precursor ions with a lifetime of 10^{-6} s or less which is the average residence time in the ion source. Fragment ions consequently appear in the mass spectrum at electron energies greater than those threshold values called appearance energies $AE(A^+)$ or $AE(Z^{+\cdot})$. In the diagram shown in Fig. 3, at higher $M^{+\cdot}$ energies $k(M^{+\cdot} \rightarrow A^+)$ becomes greater than $k(M^{+\cdot} \rightarrow Z^{+\cdot})$, so that the formation of A^+ is favoured.

The actual ion abundances in a mass spectrum are determined by both the P(E) and the k(E) relationships. The relative proportion of molecular ions in

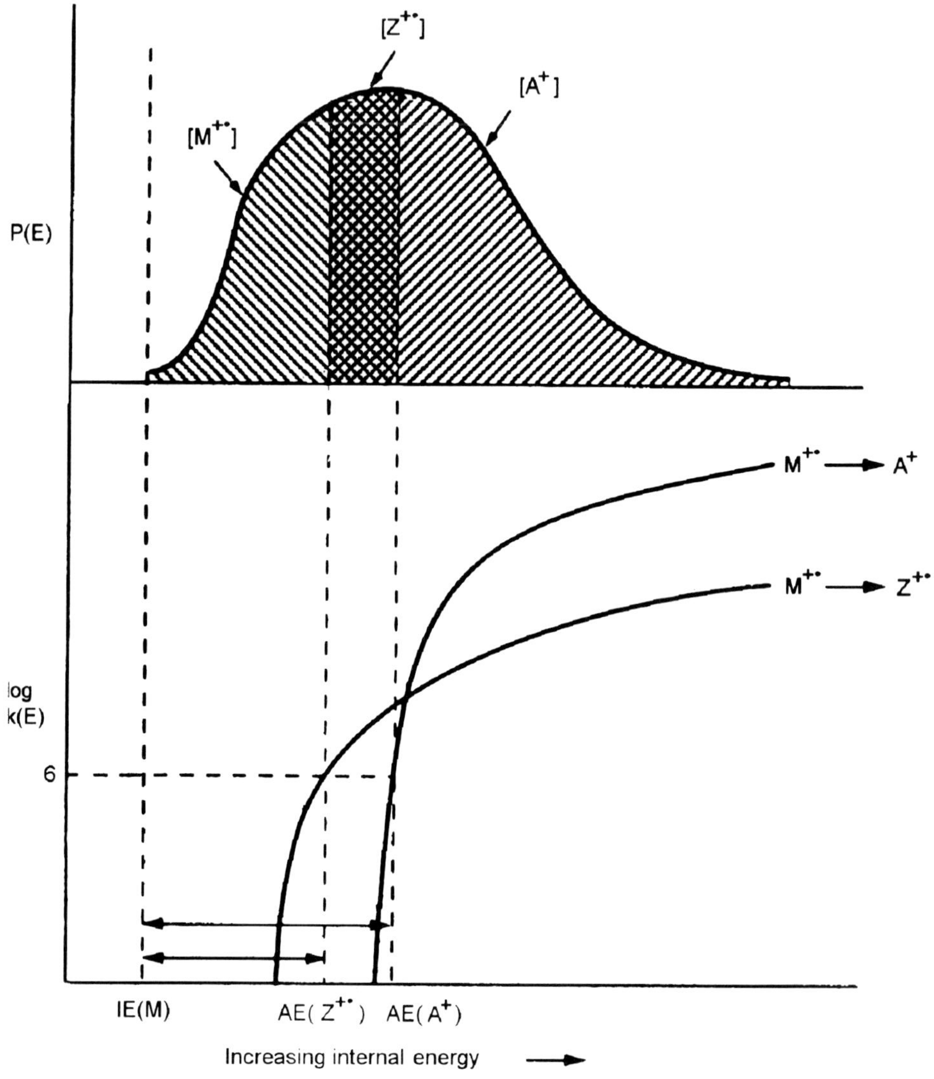

Fig. 3. Appearance of ions in a mass spectrum as a function of internal energy distribution: Relationship of probability function P(E) and rate constant k. (Modified Wahrhaftig diagram, Ref. [1])

a mass spectrum depends on the lowest energy required for ion source dissociation and the relative proportion of $M^{+\cdot}$ ions which have internal energies below this energy.

According to the quasi-equilibrium theory, ionization of the molecule, which takes place in approximately 10^{-16} s initially yields an excited molecular ion which by rapid transition between all possible energy states reaches that "quasi-equilibrium" before the various possible decompositions take place. Therefore, an ion's decomposition only depends on its internal energy and its k(E) functions. The many competing primary reactions and the consecutive secondary reactions result in a mass spectrum that can consist of a great number of peaks. The competitive nature of all these reactions (as presented for only two primary reactions in Fig. 3) is the reason that relatively small changes in molecular structure result in large differences in peak abundances.

4 Structural Information

Although the structural identification of a pesticide in a pesticide residue analysis is usually performed by means of a reference spectrum of a standard compound, a basic understanding of the interpretation of mass spectral data is certainly necessary to produce sound results.

4.1 The Molecular Ion

In a mass spectrum, several major kinds of general structural information are available. The molecular weight is probably the most valuable piece of information a mass spectrum can give. The molecular weight is calculated from the integer masses of the most abundant isotopes of the atoms present in the molecule and thus in the molecular ion. The molecular ion, also called the parent ion, is the peak that usually corresponds with the highest mass isotope cluster in the spectrum. However, identifying that peak with certainty can be rather difficult in some cases. In such cases soft ionization methods can be applied to produce ions indicative of the molecular weight. In pesticide residue analysis, chemical ionization is the soft ionization method mostly used for the production of "quasi-molecular" ions as the protonated molecule $[M + H]^+$ frequently is referred to.

In an EI mass spectrum, the fragment ions should be consistent with the molecular ion; peaks like $[M - 1]^+$, $[M - 15]^+$, $[M - 18]^+$ and $[M - 20]^+$ confirm the assignment of the molecular ion because they represent the losses of $H^{\cdot}$, $CH_3^{\cdot}$, H_2O and HF, respectively from the parent ion. Such "small-neutral loss" peaks are of major significance in deducing the molecular structure. Lists of common small, neutral fragments lost in the formation spectral peaks are given in textbooks on mass spectrometry [1, 2]. On the other hand losses of 3–14 amu and 21–25 amu generally are not consistent with fragments formed from the parent ion and indicate an incorrect assignment or the presence of impurities. With mass spectra obtained with GC-MS or LC-MS analysis, the recognition of ions originating from impurities (background or not resolved chromatographic peaks) is generally facilitated by observing the relative ion abundances in the chromatographic peak. This method is also used in an automated form as background subtraction.

As well as the molecular ion, two other types of peaks are observed in the mass spectrum: fragment and isotopic peaks.

4.2 Isotopic Peaks

The isotopic peaks are the result of neutral isotope abundances of the individual elements which can be highly indicative. For instance, natural chlorine exists as 75% 35 Cl-isotopes and 25% 37 Cl-isotopes and consequently each parent or fragment ion containing chlorine can be easily identified by its typical chlorine isotopic cluster. The abundances of isotopic peaks at unit resolution from elements occurring in pesticides are given in Table 1. There are monoisotopic elements such as fluorine, iodine, phosphorus and also hydrogen which are referred to as "A"-elements and others with typical additional isotopic peaks in the spectrum such as chlorine, bromine and sulfur that arise at two mass units higher and therefore and designated as "A + 2" elements. Molecular and fragment ions containing more than one chlorine or bromine atom therefore give rise to very characteristic patterns as shown in Fig. 4.

The isotope patterns to be expected from any combination of elements can readily be calculated and provide a useful test of ion composition. Furthermore, in compounds containing C, H, O and the heteroatoms listed in Table 1 the molecular weight must be even. Duly if a molecule contains one or an odd number of nitrogen atoms the molecular weight will be odd. This generalization applies to all stable even-electron molecules. ("Nitrogen rule").

Table 1. Natural isotopic abundances of common elements

Element	Mass	%	Mass	%	Mass	%	Type
H	1	100	2	0.015			A
C	12	100	13	1.1			A + 1
N	14	100	15	0.37			A + 1
O	16	100	17	0.04	18	0.20	A + 2
F	19	100					A
Si	28	100	29	5.1	30	3.4	A + 2
P	31	100					A
S	32	100	33	0.79	34	4.4	A + 2
Cl	35	100			37	32.0	A + 2
Br	79	100			81	97.3	A + 2
I	127	100					A

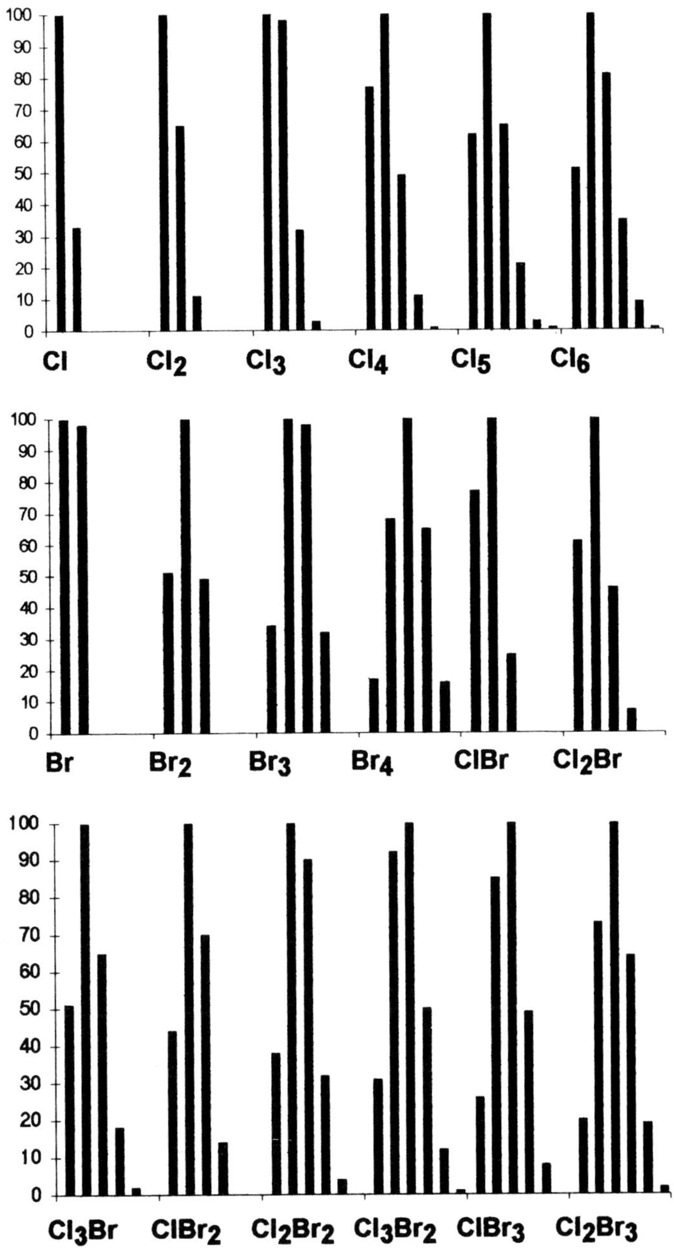

Fig. 4. Isotopic clusters

4.3 Fragmentation Reactions

Molecular ions are generated in the EI ion source with a wide range of internal energies with a significant proportion being above the threshold for fragmentation. Mass spectral reactions are unimolecular; the sample pressure in the EI ion source is kept sufficiently low to avoid collision reactions. The mass spectrum reflects the results of a series of competing and consecutive reactions occurring in the ion source as described already from the thermodynamic point of view. The reactions are thought to be initiated at the favored site for the unpaired electron and for the positive charge in the parent ion. The most favored radical and charge site in the molecular ion is assumed to arise from loss of the molecule's electron of lowest energy. Favorability for ionization generally is in the order of $\sigma < \pi <$ n-electrons from sigma bonds, double bonds (olefinic or phenyl) or non-bonding electron pairs, respectively.

Sigma-bond dissociation is typical for alkane fragmentation. The electron lost in the ionization comes from a saturated bond. The more-abundant fragment will be the one better able to stabilize the positive charge. In unbranched alkanes, the sigma-bonds are nearly equivalent in bond strength. The resulting mass spectra can have many peaks of regularly varying abundances with only a small molecular ion which often cannot be identified. The alkane spectra are easily recognized by their typical ion series with mass differences of 14 (CH_2) and therefore were called "picket fence" spectra. They were observed in nearly all chromatograms from environmental samples.

$$R{-}CH_2{-}R' \rightarrow R^{\cdot} + {}^{+}CH_2{-}R'$$

$$R{-}CH_2{-}R' \rightarrow R{-}\dot{C}H_2^{\cdot} + R'$$

Reaction initiation at the radical site arises from its strong tendency for electron pairing. The electron is donated to form a new bond to an adjacent atom. This is accompanied by cleavage of another bond. This reaction is commonly called "α-cleavage". A well-known example is the allylic cleavages with the electron lost in the ionization from an allylic double bond

$$R{-}CH_2{-}CH^{+\cdot}{-}CH_2 \rightarrow R^{\cdot} + CH_2{=}CH{-}CH_2^{+}$$

Another characteristic example is the formation of the benzylium or tropylium ion from alkyl substituted aromatic compounds.

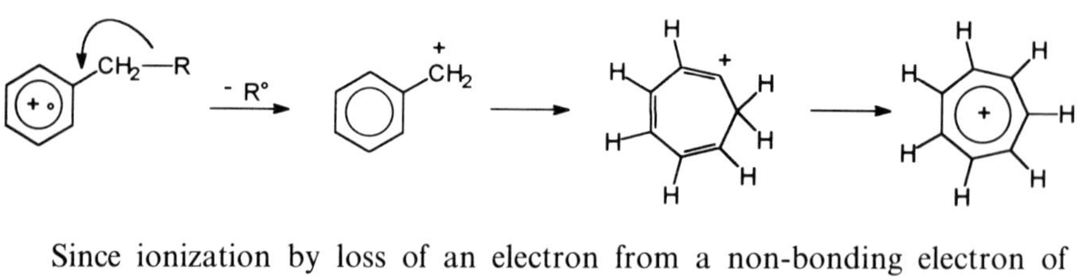

Since ionization by loss of an electron from a non-bonding electron of a hetero-atom is favored, fragments resulting from such cleavage reactions are abundant. The cleavage reaction is initiated by the positive charge which attracts an electron pair. The tendency for the formation of R^+ from R–Y parallels the inductive effect of Y. Therefore it is called "inductive cleavage" with *halogens* $>$ O, S $\gg$ N.

As well as simple bond cleavage reactions, *rearrangement reactions* are observed. In particular hydrogen atom rearrangements initiated at a radical site is an important class of reactions. Such hydrogen rearrangements through six-membered ring intermediates are usually referred to as the "*McLafferty-rearrangement*". For compounds containing an unsaturated functionality such as the carbonyl group, the γ-hydrogen atom is transferred by a sterically favorable transition state.

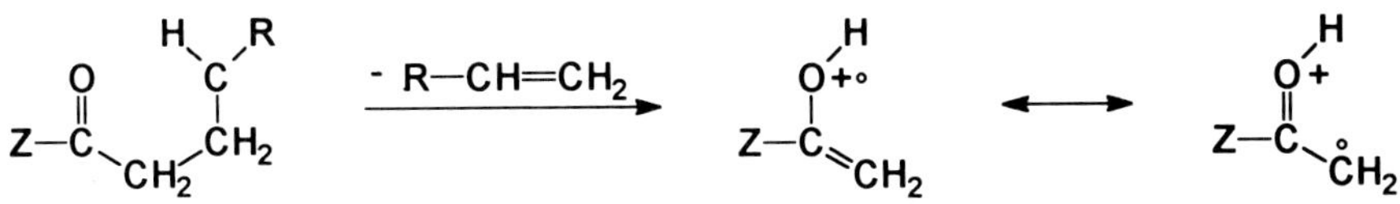

Summarizing, the most important factor affecting the abundance of a product ion is its stability which is caused by resonance stabilization (allyl or benzyl cation) or electron sharing involving a nonbinding orbital of a hetero atom, such as in an acyl ion $R{-}C^{+\cdot}{=}O \leftrightarrow R{-}C{=}O^{+\cdot}$. Another important driving force is the formation of small stable neutral products such as H_2O, C_2H_4, CO, CH_3OH, HCl and CO_2 with production of a more stabilized radical ion. The significance of small neutral loss peaks for the identification of the molecular ion in a mass spectrum has already been emphasized.

4.4 Interpretation

The principles reviewed are now illustrated with a few examples. Chlorinated pesticides are presented in Figs. 5–7 in order to demonstrate how to apply the basic knowledge to check the mass spectra and their assignment to a chemical structure for plausibility. Let us start with the methylesters of 2,4-D and dichlorprop, two herbicides of very similar chemical structure; dichorprop is derived from 2,4-D by simply substituting one hydrogen in the side chain by a methyl group. Both compounds exhibit abundant molecular ions with that of diclorprop 14 amu higher than that of 2,4-D as expected. The initially formed molecular ions are sufficiently stable, as with many aromatic compounds observed, to yield high abundances; they are the second largest peaks in the spectra. Both molecular ions exhibit even numbers and show the characteristic isotopic cluster that indicate compounds containing two chlorine atoms in their molecules as can be drawn from Fig. 4. The fragments with the highest mass are $[M-35]^+$ with 2,4-D and $[M-59]^+$ with dichlorprop representing the loss of chlorine from 2,4-D and a methoxycarbonyl from dichlorprop both easily to

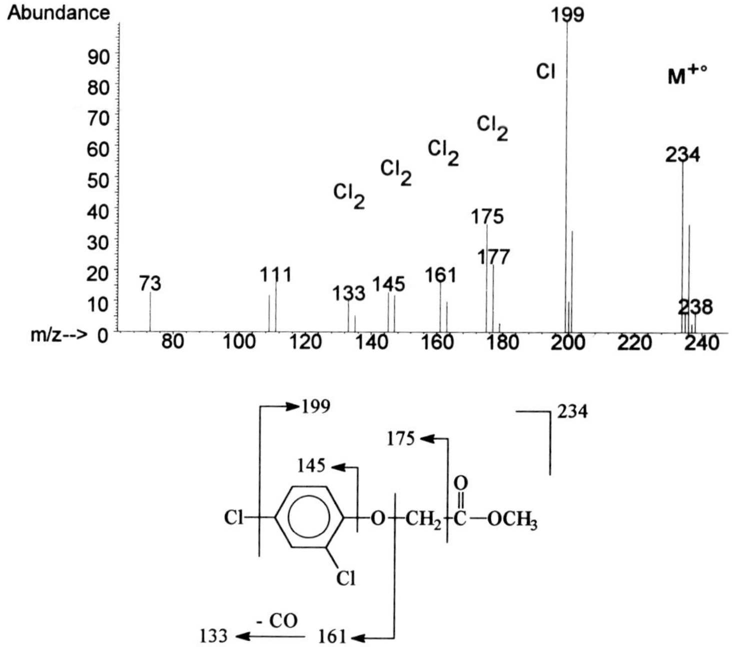

Fig. 5. Mass spectrum of the methylester of 2,4-D and the fragmentation pattern

interpret, they confirm the molecular ions. No small fragment losses are observed. The base peak m/z 199 in 2,4-D exhibits the expected isotopic cluster of one chlorine confirming the formation of that fragment by loss of one chlorine: $[M - 35]^+$. Note that the other fragments all contain two chlorine atoms and therefore support the fragmentation pattern given in Fig. 5. This also holds true for the fragmentation pattern given for the mass spectrum of dichlorprop in Fig. 6.

The base peak in this spectrum is m/z 162 which can only be explained by a hydrogen rearrangement. The formation of the peak may include a radical site rearrangement with transference of a hydrogen from the branched methyl group to the phenolic oxygen. Note the different fragmentation pattern in the two similar molecules; the dichlorophenolic ion at m/z 161 is formed from both compounds with low abundances by inductive cleavage. The possibility of undergoing a rearrangement with hydrogen migration, however, makes the formation of the diclorophenol ion at m/z 162 so favorable that it constitutes the base peak in the mass spectrum. The small difference in the chemical structure between two closely related compounds leads to mass spectra of completely different appearance. Thus these two "homologous" pesticides back up the statement that mass spectra are indicative of individual compounds and can reflect small changes in the chemical structures.

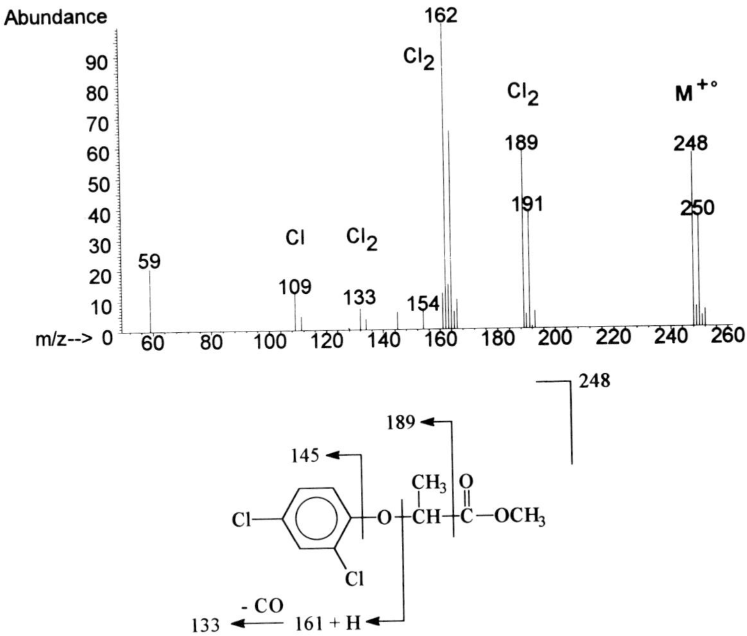

Fig. 6. Mass spectrum of the methylester of dichlorprop and the fragmentation pattern

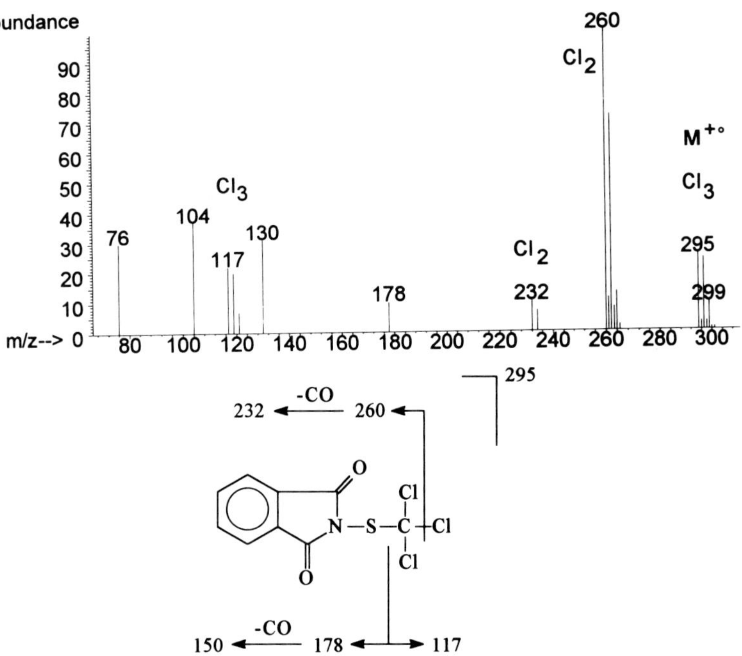

Fig. 7. Mass spectrum of folpet and fragmentation pattern

The third example shown is the mass spectrum of folpet, a chlorinated fungicide with a chlorine substituted in a methylthio group (Fig. 7). The active compound contains the heteroatoms nitrogen and sulfur in addition to the three chlorines. With only one nitrogen in the molecule the molecular ion has an odd number and is observed at a relative abundance of 30% due to its aromatic structure. The base peak ion is formed by a favorable loss of one single chlorine from the trichloromethylthio-group by inductive cleavage as expected. Only a small proportion in this reaction retains the positive charge at the trichloromethyl part (m/z 117). There is only one other chlorine containing fragment ion at m/z 232 which is formed by small neutral loss of CO. Other fragments can be explained by successive decomposition of the thiophthalimide moiety as partly indicated. Fragments retaining the aromatic acid structure are $C_6H_4CO^+$ at m/z 104 and $C_6H_4^+$ at m/z 76 being characteristic for a substituted benzene ring. The only fragment difficult to interpret is that at m/z 130 which must be formed through a complex rearrangement from the phthalimide moiety. The interpretation seems to be plausible with the chemical structure because all other major peaks can be arranged in a consistent fragmentation pattern.

5 Chemical Ionization

It became clear in the description of the basics of mass spectrometry and the interpretation of EI mass spectra that with several compounds the structural information is limited. In particular the molecular weight should be unequivocally determined.

Chemical ionization (CI) is the method of choice that can be easily applied using the mass spectrometer as a chromatographic detector. In thermospray LC-MS CI is the main ionization technique (see chapter 11). CI is very useful, in that most molecules that do not yield molecular ions by EI can produce ions with CI indicative of the molecular weight. Furthermore, CI conditions produce abundant thermal electrons that form highly efficiently negative ions from electronegative molecules by electron capture, a process familiar to the pesticide residue analyst from the electron capture detector.

For CI a reagent gas is introduced into the ion source at a concentration in large excess to that of the analytes ($10^4:1$). The reagent gas is usually ionized by electron bombardment as in EI. The formation of primary ions is followed by ion molecule reactions between those primary ions and the gas neutrals producing the chemical ionization reagent ion or a variety of such reagent ions as well as the thermal electrons.

5.1 Positive Ions

Methane is employed for chemical ionization as reagent gas most frequently because almost all organic molecules are ionized. The reactive species are formed by the following reactions:

$$CH_4 + e \rightarrow CH_4^{+\cdot}, CH_3^{+}, CH_2^{+\cdot} \text{ etc.}$$

$$CH_4^{+\cdot} + CH_4 \rightarrow CH_5^{+} + CH_3^{\cdot}$$

$$CH_3^{+} + CH_4 \rightarrow C_2H_5^{+} + H_2$$

$$CH_2^{+\cdot} + CH_4 \rightarrow C_2H_3^{+} + H_2 + H^{\cdot}$$

$$C_2H_3^{+} + CH_4 \rightarrow C_3H_5^{+} + H_2$$

At pressures around 1 torr, more than 90 percent of the ion population consists of the ions CH_5^+, $C_2H_5^+$ and $C_3H_5^+$ with m/z 17, 29 and 41. CH_5^+ reacts exothermically with almost all organic molecules behaving as a Brönsted acid to yield a protonated molecular ion:

$$M + CH_5^{+} \rightarrow [M + H]^{+} + CH_4$$

Other reactions that can be observed are:

$$M + CH_5^{+} \rightarrow [M - H]^{+} + CH_4 + H_2$$

$$M + C_2H_5^{+} \rightarrow [M + C_2H_5]^{+}$$

$$M + C_3H_5^{+} \rightarrow [M + C_3H_5]^{+}$$

The latter two equations show bimolecular association reactions which are generally classified as solvation processes in the gas phase.

Other popular reagent gases are iso-butane and ammonia which are "softer" reagent gases because they do not ionize all organic molecules and induce less fragmentation. When chemical ionization occurs by proton transfer to the analyte from an acidic reagent ion, the exothermicity of the proton-transfer reaction determines the internal energy of the protonated molecular ion and hence the extent of fragmentation.

$$M + [B + H]^{+} \rightarrow [M + H]^{+} + B$$

Table 2. Proton affinity of reactant gases

Gas	Reactant Ion	Proton Affinity (kJ mol^{-1})
CH_4	CH_5^+	527
	$C_2H_5^+$	665
H_2O	H_3O^+	706
CH_3OH	$CH_3OH_2^+$	761
i-C_4H_{10}	t-$C_4H_9^+$	807
NH_3	NH_4^+	840

The exothermicity of the proton-transfer reaction is directly related to the proton affinity. From a series of proton affinities as shown in Table 2 for a few reagent gases, the appearance of CI mass spectra is roughly predictable.

For instance, a protonated molecular ion produced via isobutane ionization is expected to possess less internal energy than that formed with methane. The series of CI mass spectra shown for two organophosphorus insecticides in Fig. 8 prove the possibility of "tailoring" the CI mass spectra to suit a particular application. For the confirmation of the molecular weight of the two pesticides methane is obviously less suited than isobutane or methanol but in a screening analysis methane may be preferred because of greater ionization yield and the production of an additional indicative fragment ion.

The suitability of CI-MS for the confirmation of the molecular weight of an analyte has two reasons. It appears that more than about 400 kJ mol^{-1} of internal excitation in the "quasimolecular ion" $[M + H]^+$ is uncommon, even

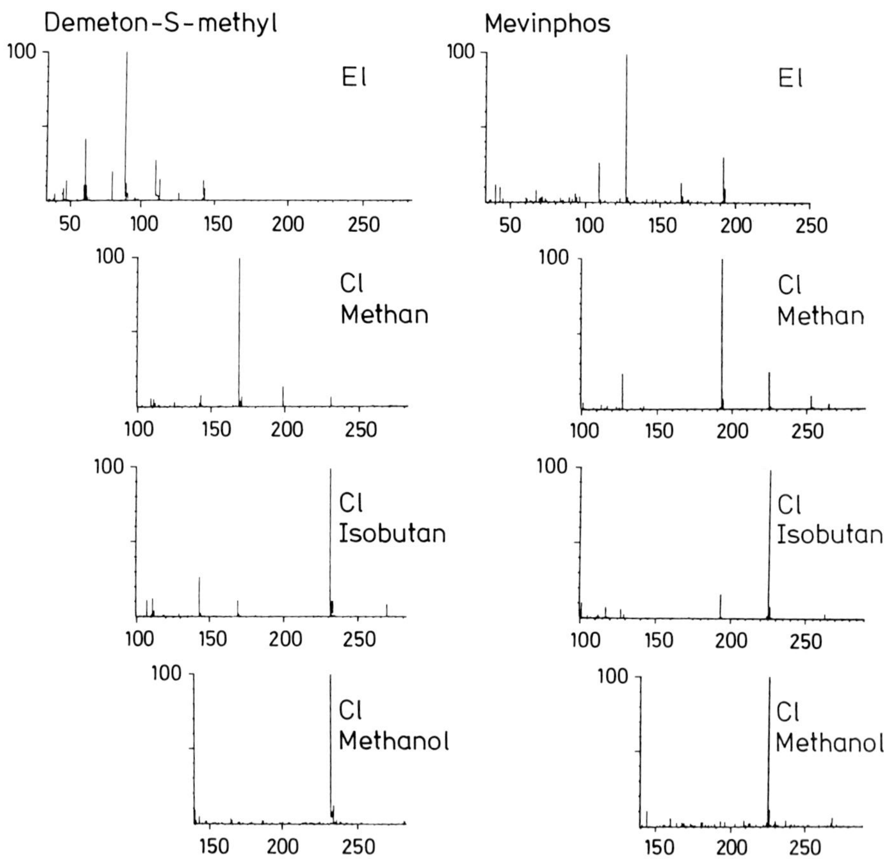

Fig. 8. PCI mass spectra of demeton–S–methyl and mevinphos (from Ref. [3] with permission)

when methane is used as the reagent gas. The consequence is that relative abundant ions appear in the molecular ion region. With labile molecules the exothermicity of the ionization can be reduced by selecting a softer reagent gas. In addition, the even-electron $[M + H]^+$ ions possess an inherent stability compared with the radical $M^{+\cdot}$ ions produced with EI.

5.2 Negative Ions

The highly energetic electrons emitted from a filament generate under EI conditions only small abundances of negative ions. Under CI conditions, however, they lose energy by promoting positive ion formation and by colliding with neutral gas molecules. The low-energy electrons produced can interact with a sample molecule AB by three different mechanisms:

$$AB + e \rightarrow AB^{-\cdot} \qquad \text{(resonance capture)}$$

$$AB + e \rightarrow A^{\cdot} + B^{-} \qquad \text{(dissociative resonance capture)}$$

$$AB + e \rightarrow A^{+} + B^{-} + e \qquad \text{(ion-pair production)}$$

"Near-thermal" electrons of very low energy (~ 0 eV) can undergo resonance capture assuming that AB possesses a positive electron affinity. With an additional large cross section for electron capture of AB such negative-ion spectra can exhibit an increase in sensitivity of orders of magnitude above those found with other ionization techniques. However, it should be noted that the extraordinary sensitivities can only be achieved under most favorable conditions with respect to the chemical structure of the analyte molecule. On the other hand, this source of information is available with all CI measurements provided the instrument is capable to detect negative ions. The most favorable equipment allows the alternate measurement of positive and negative ions in one chromatographic analysis virtually simultaneously.

Electron capture NCI (ECNCI) generates negative molecular ion radicals with low internal energy but with the inherent instability of an odd-electron ion. Therefore, the abundance of the molecular ion depends on the overall resonance stabilization possibilities in the molecule. Often an abundant stable anion constitutes the whole mass spectrum. In contrast to ECNCI, negative CI may be performed by applying special reagent gases. Brönsted bases play a role analogous to that played by Brönsted acid in generating positive ions.

For instance CH_3O^- can act as a Brönsted base, producing $[M - H]^-$ ions by abstracting a proton from the sample molecule. The ionization technique generates even-electron molecular ions of low internal energy with little fragmentation tendency. Therefore, the $[M - H]^-$ quasi-molecular ion frequently constitutes the base peak.

Since the reagent gases are not as easy to handle as that for the PCI and ECNCI, the method is not very popular yet although it has obvious merits in many applications.

6 Complementary Information

The information of mass spectra obtained from the same compound with different ionization methods is of a complementary nature. With electron ionization (EI) a parent ion is formed in the ionization process possessing an inherent instability because of being a radical or odd-electron ion. Therefore, subsequent fragmentation is common providing structural information. The abundance of the molecular ion, however, may be weak. Chemical ionization (CI) is credited with providing molecular weight information because the formation of even-electron parent ions in a gas phase reaction with proton transfer in form of "quasi-molecular" ions $[M + H]^+$ is the dominating mechanism. These even-electron ions are mostly so stable that only little fragmentation is observed. The different exothermicity of the proton transfer may, however, lead to different appearance of the mass spectra as shown in Fig. 8. With methane as the reactant gas, two more ions indicative to the molecular weight, namely $[M + 29]^+$ and $[M + 41]^+$, are usually observed making the identification of the quasimolecular ion even more reliable. With ECNCI negative odd-electron molecular ions are formed with subsequent fragmentation which frequently result in simple mass spectra dominated by a few ions. With organophosphorus pesticides often only one fragment ion originating from the organophosphate group is observed indicative to the subclass of organophosphates as diethyl- or dimethyldithiophosphate [4, 5].

To demonstrate the complementary nature of the information provided by the three ionization methods, two examples are chosen from the organophosphate insecticide, the pesticide group systematically investigated [4, 5]. In Fig. 9, the three spectra of dicrotophos are presented. The EI spectrum is dominated by the base peak indicative to dimethylphosphates which show either m/z 93 ($(CH_3O)_2PO$) or m/z 127 ($(CH_3O)_2PO_2 + H$) if hydrogen rearrangement is favoured with the side chain as is formed with dicrotophos. The molecular ion at m/z 237 and a fragment ion m/z 193 ($M-(CH_3)_2N$) exhibit low abundances. In trace analysis, these two ions may not reliably be observed. With PCI, the mass spectrum exhibits an intense "quasi-molecular" ion $[M + H]^+$ and the two adduct ions $[M + 29]^+$ and $[M + 41]^+$ at m/z 238, 266 and 278, respectively. Two fragment ions at m/z 112 and 193 are indicative for the structure of the side chain as indicated. With ECNCI, the mass spectrum of dicrotophos contains only a few negative ions with the base peak at m/z 125 dominating and indicative of the dimethylphosphate group. The fragment ion at m/z 222 stands for loss of a methyl group. All 3 fragment patterns "puzzled" together obviously give information which is sufficient to identify the pesticide as dicrotophos.

A second example is given with the 3 spectra of bromophos shown in Fig. 10. Although it belongs to the parathion group whose members usually exhibit intense molecular ions due their aromatic structure, M^+ cannot be observed with bromophos. The concentration of three halogens at one phenol moiety facilitates the expulsion of a chlorine atom radical as already seen with 2,4-D in

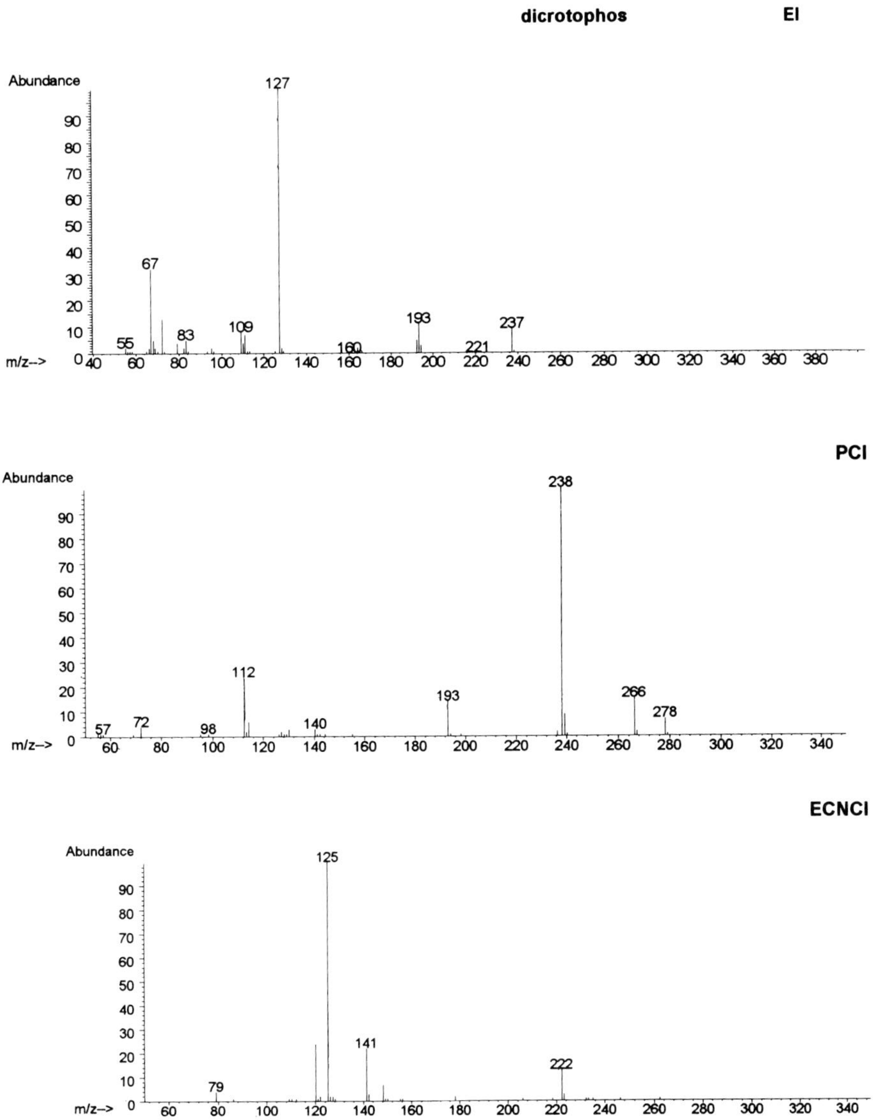

Fig. 9. Mass spectra of dicrotophos measured with different ionization conditions. *Top*: EI – *middle*: PCI– *bottom*: ECNCI

Fig. 5. The M-Cl fragment constitutes the base peak in the EI spectrum and shows the typical Cl_2Br isotope cluster (see Fig. 4). The other intense fragments from m/z 125 to 63 arise all from the dimethylthiophosphate moiety with $(CH_3O)_2$ PS at 125, $(CH_3O)_2$ PO (after rearrangement!) at 109, $(CH_3O)_2P$ at 93

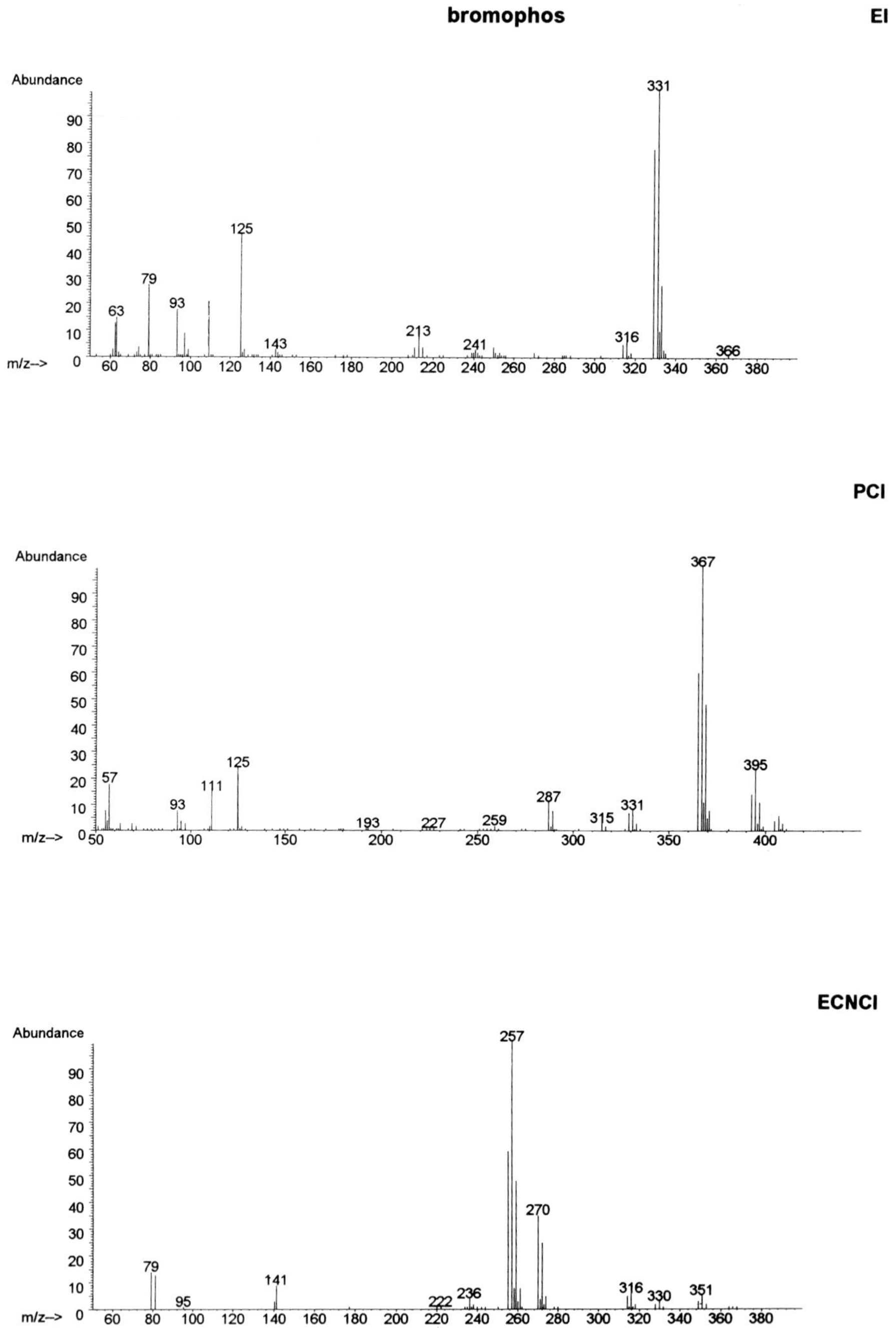

Fig. 10. Mass spectra of bromophos measured with different ionization conditions. *Top*: EI – *middle*: PCI– *bottom*: ECNCI

etc. In the PCI spectrum the $[M + H]^+$ at m/z 365/367/369/341 exhibits the expected isotopic cluster of Cl_3Br as well as the two satellite peaks $[M + C_2H_5]^+$ and $[M + C_3H_5]^+$. Loss of bromine results in m/z 287, loss of HCl in a small cluster at m/z 329 (base peak in the EI spectrum!). The fragments at m/z 125 and 111 $[(CH_3O)_2PO + 2H]$ originate from the dimethylthiophosphate group. The ECNCI spectrum consists of only two abundant ions. The base peak is formed by a rearrangement, common with the parathion group, to result in the thiophenolate with the complete halogen substitution retained as can be drawn from the isotopic cluster. The other ion is obviously formed from the M^- by loss of bromine and methyl as can be deduced from the m/z 270 and the isotopic cluster. Noticeable is the tiny peak at m/z 141 indicative of the dimethylthiophosphate group. This fragment is usually base peak with other members of this class not belonging to the parathion group but containing an alkyl side chain comparable to dicrotophos, for example. Summarizing, the molecular weight information is provided with PCI, the assignment to the dimethylthiophosphates with EI and the affiliation to the parathion group with ECNCI. The presence of two chlorine and one bromine atom is indicated by the isotopic clusters of the ions formed with all three ionization methods.

7 High Resolution Mass Spectrometry (HRMS)

HRMS enables the measurement of the exact mass of an ion and thus an unequivocal identification of its elemental composition. This high resolution can be achieved with double focusing mass spectrometers of FT-ICR instruments. The usefulness of elemental composition information increases with increasing mass, this requires also an increase in mass-measuring accuracy. The technique is of great importance in basic research when unknown chemical structures are to be elucidated. In environmental analysis, the combination GC-HRMS is applied to ultra-trace analysis of polychlorinated dioxins (PCDDs). In Table 3,

Table 3. Application of HRMS for separation of possible interferences of 2,3,7,8-TCDD

Compound	Formula	Mass of interfering ion	Resolution needed
2,3,7,8-TCDD	$C_{12}H_4O_2Cl_4$	321.8936	
Heptachlorobiphenyl	$C_{12}H_3Cl_7$	321.8678	13 000
Nonachlorobiphenyl	$C_{12}HCl_9$	321.8491	7 300
Tetrachloromethoxybiphenyl	$C_{13}H_8OCl_4$	321.9299	8 900
DDT	$C_{14}H_9Cl_5$	321.9292	9 100
DDE	$C_{14}H_8Cl_4$	321.9292	9 100

possible interferences for the most important chlorinated dioxin, namely 2,3,7,8-tetrachlorodibenzo-*p*-dioxin (2,3,7,8-TCDD) are compiled. When using the common low resolution mass spectrometry (LRMS) with GC coupling, it would not be possible to distinguish 2,3,7,8-TCDD from the other compounds listed in Table 3, if the principal ion m/z 322 was monitored in SIM mode [6].

When applying LRMS it is definitely necessary to resolve these interferences by chromatography. Otherwise, if any of these compounds co-elutes with 2,3,7,8-TCDD, false positive identification may be obtained. The other problem is to avoid false negative results that may arise from the co-eluting interference that makes the mass to intensity ratio of the indicative ions m/z 320, 322 and 324 incorrect.

Since the residue analysis of PCDDs is carried out to monitor femtogram amounts, GC-MS in SIM mode is the only method to achieve the detection sensitivity needed. Positive or negative results cannot be proved by another analytical technique and the results are of great public concern. This is the reason why official methods demand the application of GC-HRMS.

8 Tandem Mass Spectrometry (MS/MS)

The structural information provided by mass spectrometry can be further enhanced with the combination of two mass spectrometers in one instrument adding a new dimension. This technique allows the measurement of the fragmentation of a selected peak in a mass spectrum producing the daughter mass spectrum of that parent ion. The first mass spectrometer is used as a separating device for mixtures (such as unresolved peaks in GC-MS or LC-MS); after separating one particular ion, energy is added to yield dissociation product ions that are then separated in the second mass analyzer. This mass spectrum is then used for structural characterization of the parent ion. While in theoretical research this method is used to investigate the structure and stability of a molecule's fragment ions, in residue analysis the technique can be applied to the molecular species produced by soft ionization methods as CI. As described, PCI produces $[M + H]^+$ usually with little fragmentation which is useful in molecular weight determination but provides no structural information. In the most widely used instrument for MS/MS measurements three quadrupoles are combined to the so called "triple quad" with the first quadrupole as the separating device for the $[M + H]^+$ ions and the third as the mass analyzer to monitor the products of the dissociation process. This dissociation is induced by collisions with a target gas that takes place in the central quadrupole and is referred to as collision activation (CA). Such CA mass spectra are as indicative for the structure or identity of a compound as common EI mass spectra. The "triple quad" instrument can in the same way be combined with GC and LC just as a normal mass spectrometer making it the most sophisticated tool in pesticide residue analysis. Since the instrument is fully under computer control, it is possible to perform special techniques.

Tandem mass spectrometry can be carried out principally in two ways: consecutive in space by using two separate spectrometers (multiple sector or multiple quadrupole instruments) or consecutive in time by using the same mass resolving system twice (FT-IRC or ion traps). Consecutive separation in space is easier to understand and schematically shown in Fig. 11.

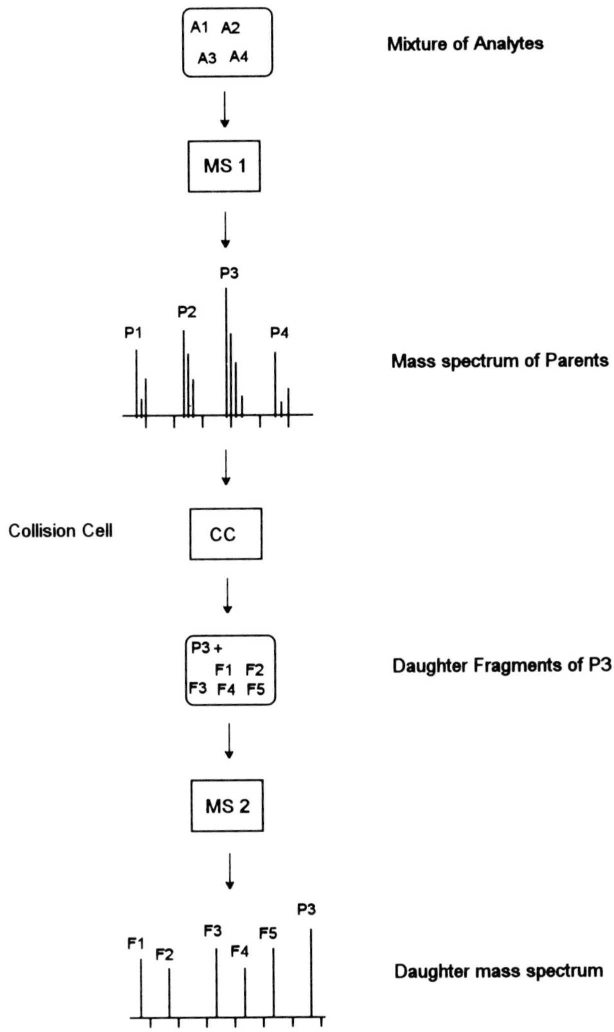

Fig. 11. Schematic of the principle of tandem MS (MS/MS)

A mixture of compounds is ionized and separated in the mass analyzer MS1. Only one parent ion is transmitted into the collision cell where the CA fragmentation takes place. The resulting fragment (daughter) ions are separated in mass analyzer MS2. Note that by selecting monoisotopic ions only monoisotopic daughter ions are produced. The daughter ion spectrum is, therefore, devoid of isotopic clusters. The basic equation describing the formation of a daughter ion m_d^+ from a parent ion m_p^+ by loss of a neutral m_n is:

$$m_p^+ \rightarrow m_d^+ + m_n$$

The parent ion is selected in MS 1, the daughter ions are detected in MS2, m_n is inferred from the difference. Each of these three species can be designated as the independent variable in MS/MS measurements, the relationships are summarized in Table 4.

The most simple reaction in MS/MS is the dissociation of a parent ion into daughter ions and neutral fragments. This is reffered to as daughter scan and provides with a full daughter ion spectrum the greatest information. To achieve lowest possible detection limits for a target compound in daughter ion MS/MS, single-reaction monitoring (SRM) is performed. In this technique all variables, as shown in Table 4 are fixed. This means that the first mass analyzer is set to transmit the parent ion and the second mass analyzer is set to transmit a specific daughter ion. This SRM technique is analogous to single-ion monitoring used in GC/MS when asking for the lowest possible detection limit in SIM mode. SRM in GC-MS/MS is a means of eliminating "chemical noise" in the MS/MS spectrum. An example is given for HPLC-MS/MS in chapter [redacted]. Although it is apparent that SRM offers less information than a complete MS/MS spectrum, it nonetheless provides a considerable increase in specificity over a single state mass spectrometric analysis. Notice that the generation of a peak in a GC-MS/MS analysis in SRM mode requires that three independent criteria be met: retention time, mass of the parent ion and mass of the daughter ion must satisfy the selected values.

A unique analytical aspect of MS/MS is the ability to screen rapidly for compound classes. Referring to the basic MS/MS reaction, any of the three species can be designated as the independent variable in an MS/MS experiment as summarized in Table 4. If the daughter ion, m_d^+, or the neutral fragment, m_n, is specified as the independent variable, new information is available that is provided by no other analytical technique.

Table 4. Parameter setting in MS/MS measurements

Scan	m_p^+	m_d^+	m_n
Daughter ion scan	fix	vary	vary
Parent ion scan	vary	fix	vary
Neutral loss scan	vary	vary	fix
Single-reaction monitoring	fix	fix	fix

Measurements in which m_d^+ is the independent variable are known as parent ion scans. If a compound class includes a particular substructure which forms a very stable ion, parent ions derived from members of this class tend to produce a common daughter ion with this substructure. With MS2 fixed to that daughter ion and MS1 scanned, all parent ions can be detected.

Measurements in which m_n is the independent variable are referred to as neutral loss scans. Many ions dissociate by loss of a small neutral molecule associated with a functional group or substructure of the parent ion. *N*-Methylcarbamate insecticides are generally characterized by loss of methyl isocyanate (m/z 57) and regeneration of the phenol which was originally used for the synthesis of the carbamate by reaction with methyl isocyanate. With a linked scan of both mass analyzers with a fixed mass difference of 57 all the phenols in a complex mixture can be selectively detected. In a neutral loss scan the isotopic cluster is fully retained, therefore the phenols can further characterized with their number of chlorine atoms.

9 References

1. McLafferty FW, Turecek F (1993) Interpretation of mass spectra, University Science Books, Mill Valley, CA.
2. McLafferty FW (1980) Interpretation of mass spectra, University Science Books, Mill Valley, CA.
3. Stan H-J (1977) Z. Analyt. Chem. 287: 104
4. Stan H-J Kellner G (1982) Biomed. Mass Spectrom. 9: 483
5. Stan H-J, Kellner G (1989) Biomed. environ. mass spectrom. 18: 645
6. Karasek FW, Clement RE (1988) Basic gas chromatography – Mass spectrometry: Principles and techniques, Elsevier, Amsterdam
7. Busch KL, Glish GL, McLuckey SA (1988) Mass spectrometry/Mass spectrometry: Techniques and applications of Tandem mass spectrometry, VCH, Weinheim, FRG

10 Suggested Reading

10.1 General Mass Spectrometry and Instrumentation

Biller JE, Biemann K (1974) Anal. Lett. 7: 515

Brunnee C (1987) Int. J. Mass Spectrom. Ion Proc. 76: 125

Busch KL, Glish GL, McLuckey SA (1988) Mass spectrometry/Mass spectrometry: Techniques and applications of tandem mass spectrometry, VCH, Weinheim, FRG

Chapman JR (1986) Practical organic mass spectrometry, Wiley, London, UK

Harrison AG (1992) Chemical Ionization Mass Spectrometry, CRC Press, Boca Raton, FL.

Howe I, Williams DH, Borden RD (1981) Mass Spectrometry: Principles and Applications, McGraw-Hill, New York
Karasek FW, Clement RE (1988) Basic Gas Chromatography – Mass spectrometry: Principles and Techniques, Elsevier, Amsterdam
March RE, Hughes RJ (1989) Quadrupole Storage Mass Spectrometry, Wiley, New York
McLafferty FW, Turecek F (1993) Interpretation of Mass Spectra, University Science Books, Mill Valley, CA.
McLafferty FW (1980) Interpretation of Mass Spectra, University Science Books, Mill Valley, CA.
McLafferty FW (Ed.) (1983) Tandem Mass Spectrometry, Wiley, New York
Middleditch BS (ed) (1977) Practical Mass Spectrometry: A Contemporary Introduction, Pergamon, Oxford, UK

10.2 Application to Environmental Analysis

Barcelo D (1992) In: Cairns Th, Sherma J (eds.) Emerging Strategies for Pesticide Analysis. Hyphenated Methods for Pesticide Residue Analysis, CRC Press, Boca Raton, FL, p. 273.
Brown MA (1990) Liquid Chromatography/Mass Spectrometry Applications in Agricultural, Pharmaceutical and Environmental Chemistry, American Chemical Society, Washington
Cairns Th, Sherma J (eds.) (1992), Emerging Strategies for Pesticide Analysis, CRC Press, Boca Raton, FL.
Karasek FW, Hutzinger O, Safe S (eds.) (1985) Mass Spectrometry in Environmental Sciences, Plenum, New York
Mattern GC, Rosen JD (1992) In: Cairns Th, Sherma J (eds.), Emerging Strategies for Pesticide Analysis: the Use of Ion Trap Mass Spectrometry for Multiresidue Pesticide Analysis, CRC Press, Boca Raton, FL. p. 259.
Rosen JD (ed.) (1987) Application of New Mass Spectrometric Techniques in Pesticide Chemistry, Wiley, New York
Shalaby LM, George SW (1992) In: Cairns Th, Sherma J (eds.) Emerging Strategies for Pesticide Analysis: LC/MS Multiresidue Methode for Thermally Labile Pesticides, CRC Press, Boca Raton, FL, p. 295

10.3 Mass Spectral Libraries

Eight Peak Index of Mass Spectra, Mass Spectrometry Data Centre Royal Society of Chemistry, UK (1991)
McLafferty FW, Stauffer DB (1989) The Wiley/NBS Registry of Mass Spectral Data, Wiley, London, UK
McLafferty FW, Stauffer DB (1991) Important Peak Index of the Resgistry of Mass Spectral Data, Wiley, London, UK

10.4 Mass Spectral Libraries Dedicated to Environmental Analysis

Cairns, Th., Siegmund EG, Jacobson RA (eds) (1987) Mass Spectral Data Compilation of Pesticides and Industrial Chemicals, Food and Drug Administration, Los Angeles, CA

Freudenthal J, Gramberg LG (1975) Catalogue of Mass Spectra of Pesticides, National Institute of Public Health, Bilthoven, NL
Hites RH (1985) Handbook of Mass Spectra of Environmental Contaminants, CRC Press, Boca Raton, FL.
NIST/EPA/MSDC Mass Spectral Database, The National Institute of Science and Technology, Gaithersburg, MD. (1990)
Stan H-J, Lipinski J (1989) HP Pesticide Library, Hewlett Packard, Palo Alto, CA
Waller GR (ed.) (1972) Biochemical Applications of Mass Spectrometry, Wiley, New York
Waller GR, Dermer OC (eds) (1980) Biochemical Applications of Mass Spectrometry, Wiley, New York

Identification and Confirmatory Analysis Based on Capillary GC-Mass Spectrometry

Hans-Jürgen Stan and Thomas Heberer

Institute of Food Chemistry, Technical University Berlin, D-13355 Berlin, FRG

In this chapter, the application of GC/MS for the confirmation of results positively monitored with screening methods using less selective detectors is described. Furthermore, the screening for pesticides in water–ranging from ground and tap to surface and sewage water–with full scan as well as the target compound analysis applying selected ion monitoring (SIM) is reviewed. The chapter starts with a short description of the different types of mass spectrometric equipment with respect to their suitability for GC/MS coupling. State of the art pesticide multiresidue analysis in water is demonstrated with examples from different pesticide classes. Since the examples in this chapter were mainly produced by the authors group, a survey of recent publications was prepared and is presented in a preceding subchapter. Furthermore, a case story of the elucidation of an unknown contaminant in ground water samples is reported. Finally a software program is described enabling automatic screening of water samples for a multitude of pesticides by searching dedicated pesticide libraries with full scan acquisition

1 Introduction and Scope

Multi-residue screening procedures for several hundred pesticides in water as well as in food and other environmental samples are based on gas chromatographic determinations. Results obtained with popular selective detectors such as ECD, NPD and FPD require confirmation by GC-MS. This applies also for results obtained with the element-specific AED.

GC-MS has dominated confirmatory analysis in the pesticide field from the early days.

This technique has greatly benefited from the development of fused silica capillary columns and the development of small, relatively inexpensive mass-spectrometers as dedicated gas chromatographic detectors. GC-MS is now readily available to residue chemists and the ease of operation and maintenance make specialists in mass spectroscopy no longer a prerequisite for GC-MS operation as with the more complex instruments of earlier generations. Most of the examples given in this chapter were obtained with mass selective detectors.

Positive identification of low level residues in a water sample presents the analyst with a number of problems. Full scan spectra should be obtained wherever possible. The high sensitivity and selectivity of modern GC-MS instruments enables this in many situations to below 1 μg per liter depending on the matrix and in particular on the chemical structure of the pesticide. With most instruments, full scan spectra can be evaluated at the low ng level, that means 1 or 2 ng analyte injected into the GC-MS system with the sample. This requires that extracts from water samples must be adequately cleaned-up and concentrated. Spectral averaging and background subtraction facilities provided by the data system are generally used to remove contributions from background or partially resolved contaminants. However, with very weak spectra these data processing procedures may lead to corrected mass spectra of dubious validity. This is the point where the analyst has to change from full spectra scanning to selected ion monitoring using the reduced number of mass channels with the considerably lower detection limits for the specified target compounds.

The principles of mass spectrometry and how a mass spectrum emerges and can be interpreted are reviewed in chapter 5. In this chapter, the application of GC/MS for the confirmation of results positively monitored with screening methods using less selective detectors is described. Furthermore, the screening for pesticides in water with full scan as well as the target compound analysis applying selected ion monitoring (SIM) is reviewed and finally, a case story of the elucidation of an unknown contaminant in ground water samples is reported. Since the examples in this chapter were mainly produced by our group, a survey of recent publications was prepared and is presented in a preceding subchapter.

2 Application of GC-MS to Water Analysis

Gas chromatography with mass selective detection plays a major role in confirmation of results for pesticide residues in water analysis [1, 2]. Additionally, GC-MS detection is described for screening analysis of pollutants, including pesticides in drinking and surface water by EPA methods 525 and 625 [3, 4]. Using full scan acquisition, these methods provide detection limits at the ppb level for the pesticides investigated. Thus, these methods do not provide the sensitivity required by the drinking water regulations of the European Community [5], which are fixed at 0.1 ppb for drinking water analysis. Otherwise, confirmation of results obtained in screening analysis with ECD or NPD detection meeting these low concentration levels can easily be performed using GC-MS in selected ion monitoring mode. Confirmation methods using electron impact or chemical ionization in water analysis are reported for acidic herbicides [6–12], organochlorine pesticides [13], triazines [1, 2, 9, 11, 14–17], organo phosphorous pesticides [1, 2, 15, 18], carbamate and urea pesticides [19], triazoles and pyrimidines [20], pyrethrines and pyrethroides [21, 22] and sulfonylureas [23]. These methods provide detection limits sometimes far beyond the 100 ng per liter level.

As a consequence of and supported by decreasing prices of GC-MS equipment, GC-MS with SIM is already widespread for screening analysis of pesticides in water [1–4, 10, 14, 24–53]. Multi-residue methods using SIM have gained great importance for trace analysis, especially with phenoxyalkanoic acids [1, 2, 46, 51, 53]. Often this is a consequence of the necessity to derivatize polar compounds prior to gas chromatography. In this case only a limited number of compounds have to be targeted. Furthermore derivatized matrix compounds or by-products of the derivatization procedure often interfere with detection with conventional detectors. For the phenoxyalkanoic acids, methylation of the carboxylic function by diazomethane [1, 2, 6, 32, 39, 40, 45] or methanol/sulphuric acid [10, 41, 43, 46] and pentafluorobenzylation using pentafluorobenzyl bromide [7, 10, 41, 43, 46, 51–53] are most commonly used. However, methylation does not solve the problems with ECD detection of lower chlorinated analytes, while PFB-esters increase the ECD sensitivity tremendously, but provide many interference problems especially with environmental water samples [7]. Thus, the screening of these derivatives using GC-MS with SIM has great advantages compared to conventional ECD detection [1, 2]. With SIM programming the analysis of up to 35 acidic herbicides has been reported at the low nanogram per liter level [51–53].

A compilation of recent publications using GC-MS for the screening of pesticides in water analysis is presented in Table 1.

In this chapter, multi-residue SIM methods with more than 50 chlorinated pesticides, triazines and acidic herbicides will be presented. The principles of this methology and its application to environmental water samples at the nanogram per liter level will be demonstrated.

Table 1. Compilation of recent literature on screening analysis of pesticides in water samples using GC-MS

Target pesticides					No.	Detector	Ionization Technique	SIM	Ref.
CP	TP	PAA	OP	Others					
x					18	TD-CG-MD		x	24
				x	11	HP MSD 5970	electron impact	x	25
				x	13	Fin. MAT ITD 700	electron impact	x	26
					32	HP MSD 5970	electron impact	x	27
x	x		x	x	35	HP MSD 5970	electron impact	?	28
					20	ITD	chemical ionization		29
?	?	x	x	x	39	JMS-DX303	electron impact	x	30
	x		x	x	13	HP MSD 5970	electron impact	x	31
x	x	x	?	x	20	ITD	electron impact		32
x	x		x	x	21	ITD	chemical ionization	x	34
	x				9	HP MSD 5970	electron impact	?	35
	x				5	Finnegan 4535			36
x	x	x	x	x	103	CE QMD-1000	electron impact	x	37
x	x		x		50	VG TS 250	electron impact	x	38
x					15		electron impact		4
x	x			x	14		electron impact		3
		x			12	HP MS 5988A	electron impact	x	10
		x			13	HP MSD 5970	electron impact	x	39
		x			17	HP MSD 5970	electron impact	x	40
		x			8	Fin. MAT ITD 700	electron impact		41
		x			8		electron impact	(x)	42
		x			9	HP MS 5988A	electron impact	x	43
		x			14	Fin. MAT ITD 700	electron impact		44
x		x			17	HP MS 5995	electron impact	x	45
		x						x	46
				x	7	HP MSD 5970	electron impact	x	47
	x				8			x	14
		x							2
		x							1
x	x	x			55	HP MSD 5970	electron impact		50
		x			35	HP MSD 5970	electron impact	x	51
		x			35	HP MSD 5970	electron impact	x	52
		x			36	HP MSD 5970	electron impact	x	53
									33
									48

CP: chlorinated pesticides, TP: triazine pesticides, PAA: phenoxyalkanoic acids and related compounds,
OP: organophosphorous pesticides

3 The GC-MS Instrument

Capillary GC is the analytical method with the greatest separation power. MS is the most sensitive method of molecular analysis with the potential to yield information on the molecular weight as well as the structure of the analyte. When these two methods are directly combined into one GC/MS system, the capabilities of that system are not merely the sum of the capabilities of the two outstanding analytical methods; the increase in analytical information is exponential. Extreme selectivities can be obtained which are of utmost importance in screening analysis of target compounds in environmental samples. The extremely numerous data generated by the GC/MS system in one single analysis makes a dedicated computer necessary.

Automated analysis is routine in GC with environmental samples. Autosamplers carry out automatic injection in splitless or on-column or any other mode fully controlled by built-in software. The enormous amount of data generated in each GC-MS analysis is stored in data files usually on a hard disk. In this way sample throughput can be maximized by round-the-clock instrument operation. The analyst is no longer confined to instrument operation but confronted with an abundance of analytical data. In screening analysis it is highly desirable at least to select positive or possibly positive samples from those certainly free of residues of pesticides or other target compounds.

In principle, any problem that can be solved using pen and paper or a keyboard of a computer in a finite amount of time by following logical rules can be performed by a computer alone. However, it is difficult to define clearly such rules for GC-MS data analysis. The key to flexibility necessary for creating automated methods is to break down the data analysis problem into a number of small tasks, each of which has associated software routines. These routines can then be linked together as needed to customize data analysis for individual samples with respect to target compounds. Considerable improvements in the available software have been introduced in recent years for qualitative automated data analysis providing powerful macro programming language enabling links with standard text and spreadsheet programs.

4 The Mass Spectrometer

When a molecule is ionized in a vacuum, a characteristic group of ions of different masses is formed. When these ions are separated, the plot of their relative abundances versus mass constitutes a mass spectrum. The emergence of such a mass spectrum and the information that can be drawn from it is described in chapter 5.

Mass spectrometry can be divided into two fundamental processes: ionization and mass separation or filtering with subsequent recording of the ions formed. The recorded ions are finally subject to data processing by means of computers.

The mass spectrometer nowadays is a highly sophisticated instrument under full computer control. It basically consists of five parts for: sample introduction, ionization, mass analysis, ion detection and data processing.

4.1 Sample Introduction

Sample introduction in capillary GC-MS is nowadays simply performed by conducting the end of the fused silica column directly into the ion source by means of a heated transfer line. Modern mass spectrometers are equipped with efficient pumps to cope with the flow of carrier gas from the column up to 20 $ml\,min^{-1}$, values commonly encountered with wide bore capillary columns. Narrow bore columns, however, are operated usually with a carrier gas flow of less than 2 $ml\,min^{-1}$.

4.2 Ionization

The ionization of the analytes can be performed in a number of ways but for automated screening analysis only electron impact ionization is in use although special applications of target analysis with other ionization techniques are possible. The various ionization methods employed in pesticide residue analysis are reviewed in chapter 5.

4.3 Mass Analyzers

After their production in an ion source, ions are analyzed according to their mass-to-charge ratio (m/z) in a mass analyzer. Five types of mass analyzers are currently available, the magnet sector, quadrupole mass filter, ion trap, time-of-flight and ion cyclotron resonance instruments. The first three are generally applied in GC-MS systems.

4.3.1 Magnetic Sector Instruments

The magnetic sector mass spectrometer is the type on which all the early work in organic mass spectrometry was performed as well as the pioneering work in GC-MS by coupling packed columns to an ion source by means of special interfaces which preferentially removed carrier gas molecules and transferred the analyte molecules to the ion source. An electromagnet is used to separate ions for subsequent mass detection. In a single focusing sector instrument, the ions with mass m and z elementary charges are introduced into a magnetic field with a particular kinetic energy which is determined by the accelerating voltage. The ions are accelerated towards the source exit slit with a great deal of energy by means of a repeller plate and influenced by the magnetic field which focuses ions of the same mass-to-charge ration into a narrow beam which passes through a slit just prior to the detector. By variation of the magnetic field (or the accelerating voltage), ions of different m/z values can be detected by a detector at a fixed position as being separated in time. The most common way of scanning is by an exponential magnet scan allowing equal dwell times for all individual masses within the scan. The resolution of the mass analysis can be improved by means of an electrostatic analyzer which provides an additional focus to the ions. Instruments with both a magnetic and an electrostatic sector are called double focusing mass spectrometers. These are capable of high-resolution mass determination separating different ions with the same nominal masses and mainly used for elucidation of chemical structures of unknown compounds. In environmental analysis, however, they are state of the art in the trace analysis of polychlorinated dibenzodioxins and dibenzofuranes.

4.3.2 Quadrupole Instruments

The quadrupole mass analyzer actually is a mass filter. It consists of four hyperbolic rods that are placed parallel in a radial array. Opposite rods are charged by a positive or negative DC voltage upon which an oscillating radio frequency is superimposed. Ions are introduced into the quadrupole field by means of a low accelerating potential of typically 10–20 V. They start to oscillate in a plane perpendicular to the rod length. When the oscillations are not stable, the ions do not pass the filter because the amplitude of the oscillations becomes infinite. When stable trajectories are made, the ions are transmitted towards the detector. The quadrupole filter thus acts as a bond-passfilter, usually set to transmit ions of one particular m/z-ratio ('unit-mass resolution'). To obtain a mass scan, the DC and radio frequency voltages are varied while their ratio is kept constant. The mass permitted to pass through is linearly related to the amplitude of the voltage. This simplifies GC/MS operation as well as computerization. The linear relationship between mass and voltage makes control and calibration by computers easy. Quadrupole mass spectrometers have a reputation for high sensitivity and the ability to scan rapidly at millisecond intervals. These qualities made them well suited for coupling with capillary gas chromatography to scan the narrow peaks produced.

The quadrupole mass filter is the most widely applied mass analyzer in GC/MS as well as in LC/MS.

4.3.3 Ion Trap Detectors

The ion trap was developed as an quadrupole-related detector for capillary gas chromatography. The unique feature of the ion trap compared to conventional mass spectrometers is that the ion source and analyzer region are the same. Ion traps are operated at relatively high pressures (0.1 pa of He).

Molecules entering the trap are ionized by conventional electron impact. Ions over the entire m/z range of interest are not allowed to leave, they are trapped by a quadrupole field which is formed between end cap electrodes and a ring electrode by applying a radio-frequency voltage. By raising the RF potential the trajectories of ions of successive m/z values are made unstable. Unstable ions will rapidly leave the trapping field region in the direction of the end cap electrodes, and since the lower end cap is perforated, a significant percentage will be transmitted through and are detected by an electron multiplier. Detection limits reached with the ion trap detector are reported to be lower than with any other mass spectrometric detector in full scan mode but there is no increase in detection sensitivity in SIM mode that is observed with quadrupole or magnetic sector instruments. The ITD of Finnigan was originally designed as a stand-alone gas chromatographic detector that easily can be interfaced to any GC using capillary columns. Actually however, the ion trap is sold mostly as an integral part of a computerized GC/MS instrument comparable to other GC/MS instruments basing on quadrupole technology. Mass spectra generated by an ion trap are not always identical to those from conventional quadrupole mass spectrometers, although differences are generally not great. The reason is that the pressure in the ion trap is higher than in a conventional ion source forming $(M + 1)^+$ ions from addition of H^+. This ion-molecule reaction resembles production of pseudo molecular ions in chemical ionization. Therefore, under certain conditions, EI mass spectra may contain a few additional ions resulting from the chemical ionization process. Recent improvements in the computer control of the ionization process have reduced the ion-molecule reactions such that "mixed EI/Cl mass spectra" no longer are recorded under conditions of routine GC/MS analysis.

4.3.4 Time-of-Flight (TOF) Instruments

In a time-of-flight mass spectrometer a pulsed beam of ions is accelerated by a potential and the time needed to reach a detector is measured. The ion source is pulsed in a way that a full mass spectrum is recorded before the first ions of the next pulse arrive at the detector. Important characteristics of the time-of-flight instrument are high ion transmission and no upper mass limit. The resolution is limited, although impressive improvements in this respect have been achieved recently. This type of mass spectrometer is extensively used in studies with plasma and laser desorption techniques on large molecules such as proteins but not interfaced to gas chromatography.

4.3.5 Fourier-Transform-Ion Cyclotron Resonance (FT-ICR) Instruments

In a fourier-transform ion cyclotron resonance mass spectrometer (FT-ICR/MS) the mass analysis is performed in a cubic cell placed in a magnetic field. The cell consists of two opposite trapping plates, two opposite excitation plates and two opposite receiver plates. Ions are trapped in the cell in cyclotron motions and can be excited by means of a radio-frequency pulse to move them in phase on increased circle radiuses. The coherent movement of the ions generates an image current in the receiver plates that finally can be transformed by applying fourier transformation into a regular mass spectrum. An important feature of FT-ICRMS is the extremely high resolution and sensitivity that can be achieved. The cell must be placed in extreme vacuum (10^{-7} Pa). Although interfacing to a gas chromatograph has been reported; the domain of the high cost instrument is basic research rather than analyzing environmental samples.

The relative merits of the various mass analyzers are discussed in detail by Brunnee [54]. The choice of the analyzer depends on the application. In practice, most GC-MS instruments have been developed using quadrupole technology including ion traps. This choice is mainly determined by the simplicity, costs, vacuum and consequent space requirements.

4.4 Ion Detection

All mass spectrometers that are easily interfaced to a gas chromatograph are nowadays equipped with an electron multiplier. In such an electron multiplier the ion beam is converted to an electron beam that is subsequently amplified through a cascade effect-with a gain in current of ca. 10^6. In analog detection, the signal of the multiplier is further amplified and finally converted by means of an analog to digital converter into digits that can be processed by a computer. Usually the electron multiplier is constructed to detect positive ions but by placing a conversion dynode in front of the electron multiplier negative ions can be detected, too. Upon impact of negative ions the conversion dynode produces positive ions which are amplified as described.

4.5 Data Acquisition and Processing (Handling)

Modern GC-MS systems produce vast numbers of data that are acquired using fast and expensive computers. The operation of the mass spectrometer, data acquisition and processing is fully executed and controlled by the computer. Additionally, the interpretation of the data is also done by the computer to a considerable extent, especially in the form of automated library searches against reference spectra compiled in specialized libraries.

Two modes of operation are in common use and also applied to automated screening analysis of pesticide residues: Repetitive scanning also described as cyclic scanning or full scan mode and selected ion monitoring (SIM).

In full scan mode, data are acquired by continuous repetitive scanning of the GC column eluate over the full analysis time starting after the solvent peak has been passed. The rate of scanning is predetermined by the operator; usually values in the range of 0.5 or 1 s per scan are used with capillary columns. Each scan results in a full mass spectrum that is stored in the computer memory separately.

Basically, a three-dimensional data- array is generated by repetitive scanning with time, mass-to-charge ratio and ion intensity as the three dimensions. This data-array can be processed in various ways. A section in the plane of mass-to-charge ratio is called a mass chromatogram or ion chromatogram. When the intensities of all ions in each spectrum are summed and plotted as a function of time, a total ion current chromatogram is obtained (TIC). In GC/MS, this plot is used as the non-selective chromatogram to see all compounds in the sample amenable to gas chromatographic analysis. This chromatogram is often compared to those obtained with the universal flame ionization detector.

Mass or ion chromatograms are here referred to as reconstructed ion chromatograms (RIC) in order to emphasize the fact that they are produced by reconstructing a chromatogram-like ion intensity plot from separate mass spectra acquired in repetitive scanning and to distinguish them clearly from SIM.

4.6 Tuning and Calibration

The information that can be obtained from a mass spectrum fully relies on proper tuning and calibration of the instrument. Tuning is performed to achieve a high sensitivity over the whole mass scanning range and a proper mass resolution. The former tedious task of iterative adjustment to obtain the desired performance is with modern instruments automatically carried out by software algorithms which optimize the interactive lens potentials.

The calibration of the m/z axis of the mass spectrum is performed with reference compounds of which perfluorkerosine (PFK) and heptacosafluorotributyl amine (PFTBA) are in general use over the mass range relevant for GC/MS.

5 Compound Identification

5.1 Mass Spectral Libraries

Mass spectra obtained under standard conditions may be considered as a finger print of the molecule reflecting its chemical structure. They have been therefore collected in various mass spectral libraries. These libraries are available for computer searching and identification of unknown compounds provided that a clean mass spectrum can be produced with the analytical procedure. Excellent search methods for computerized libraries are available but the usefulness of these methods must not be overestimated with respect to elucidating the identity of unknown compounds in an environmental sample because only a minor portion of all the organic compounds known are compiled in the universal mass spectral collections. In target analysis as pesticide residue analysis, however, the situation is much better because all the peaks in chromatogram are compared by means of their mass spectra with the entries of a limited mass spectral library containing the target compounds.

5.2 Background Ions

A major problem in identification of sample peaks by means of their mass spectra is background ions. These may be confusing in the interpretation of mass spectra. In electron impact mass spectrometry there are background peaks in the lower mass region at m/z 18, 32, 40 and 44 due to residual air leaks. In GC/MS analysis, background ions may arise from column bleed of the separation phase and from the carrier gas. Since these background ions appear constantly over the whole gas chromatogram they can easily be eliminated by background subtraction. Additionally, impurity peaks may arise from the sample preparation and clean up, phthalates e.g. and other plasticizers are ubiquitous and practically unavoidable in trace analysis. The main problem with background ions in mass spectra of gas chromatographic peaks, however, arises from incomplete separation of the analytes from matrix compounds.

5.3 Background Subtraction

Fortunately, computer background correction enables the removal of background ions from the analyte spectrum in most cases. The easiest way to clean up mass spectra is to subtract another mass spectrum which contains only background ions. Background ions are common to a larger number of mass spectra scanned at the base line of the gas chromatogram while ions due to a sample component exhibit abundances following chromatographic peak shapes with a maximum at the apex of the peak. The simplest type of subtraction involves a mass-by-mass subtraction of ion abundances of a background spectrum from the ion abundances at corresponding masses from the mass spectrum at the apex of the peak. Improved results can sometimes be obtained by averaging two or three spectra taken across the top of the GC peak, and subtraction of averaged spectra from both the leading and the trailing edges of the peak. This method is of particular value in resolving overlapping peaks.

Automated background subtraction methods can be used to remove these interfering ions from overlapping matrix peaks as well as from non-separated target compounds. Automated background subtraction procedures are based on an algorithm which was introduced by Biller and Biemann as early as 1974 [55]. The procedure identifies all ions that maximize at each scan number and strips away all other ions from that scan. This stripping procedure not only removes background ions from the column bleed and common matrix effectively but also removes ions in a mass spectrum that originate from closely-eluting unresolved GC peaks.

Modern GC/MS instruments have such a software routine for automated background subtraction at the operator's disposal. On the other hand, other methods of background subtraction as described for manual evaluation can be easily designed using macro language facilities.

5.4 Library Search

The most important step in automated screening for pesticides or other target compounds is the possibility of automated library search. When measured under standard conditions, the mass spectrum of a molecule is very indicative (like a fingerprint). By comparing the mass spectrum of the analyte with others in a reference file, the identity of its nature can be recognized. A useful feature of all computerized comparison algorithms is the calculation of factors which are used to distinguish between good, average and poor matches. A human would generally stop searching at the first good match, but a computer is usually programmed to find all matches above a given threshold of match factor and to report them in a rank list. With modern mass spectral search software more than one match factor is reported reflecting the search algorithm applied.

The performance of a library search routine is not to check by theoretical considerations but only by application to standard mixtures and spiked samples. Analyzing standard mixtures with decreasing concentrations gives the analyst a meaning of the instruments detection sensitivity. In other words, he will learn which amount of a particular pesticide must be injected to obtain a positive identification with a full scan spectrum. The limits of detection vary with the target compounds depending on their fragmentation behaviour. This includes the abundance of the molecular ion and that of fragment ions in the high mass region as well as the presence of isotopic clusters. The detectability, however, also depends considerably on the chromatographic properties of the compound: good gas chromatography is a prerequisite of reliable results in residue analysis with GC-MS at trace level concentrations (see chapter 3).

5.5 Manual Verification – RIC – Background Subtraction

Library search results can show a poor hit quality but excellent correlation in the retention times for the peak searched and the suggested library compound. In this case, the target compound may be overlapped by a co-eluate from the matrix and manual evaluation must, therefore, be performed. Such a manual verification procedure, performed to confirm the identity of a compound generating a peak overlapped by the peak of a matrix compound was applied in the discovery of the unknown contaminant in Berlin ground water which finally was identified as clofibric acid as described in the case history in Sect 5.9.4 and chapter 8. As described, the identification of the unknown was possible having obtained a clean mass spectrum. This was the result of a careful inspection of the data and the good separation power observed with the capillary column used. The slight separation of the overlapped peaks made possible the assignment of the individual peaks constituting the mixed mass spectrum to the two components. One component was obviously one of the phthalic esters ubiquitous in environmental samples. The retention time indicated di-isobutyl phthalate. The hypothesis of two compounds overlapping in that one peak was proved by the use of reconstructed ion chromatograms (RIC). The RIC of the selected specific ions of the former unknown clofibric acid PFB ester, m/z = 128 and m/z = 394, and m/z = 149 which is indicative of the coeluting phthalate are shown in Fig. 1.

It is evident that the traces of the target compound ions and that of the ion belonging to the plasticizer show that the two compounds to not elute exactly at the same time. In order to obtain a target compound spectrum of maximum purity, a scan was taken at that time where the plasticizer (ion m/z 149) has almost completely eluted. After background subtraction, a clear mass spectrum of clofibric acid PFB ester was obtained that enabled the described elucidation of the structure. As shown in the example, manual evaluation of peaks exhibiting poor hit quality to suggested library compounds, but a good correlation in retention time can provide excellent results when using RIC of appropriate selected ions. A disadvantage of this technique is that it is very time consuming when applied to a great number of peaks and should only be necessary when identifying unknown contaminants. Therefore, the macro program AUTARG Level 2 was developed to execute automatically all the steps described above when monitoring target compounds [49, 50]. The characteristics of this program and its application are described later in this chapter.

5.6 Screening for Chlorinated Pesticides and Triazine Herbicides

Since a maximum tolerance of 100 ng l^{-1} for all pesticides was established with the EC Drinking water guidelines, this concentration level has become the standard to evaluate all analytical methods in the field of residue analysis. This

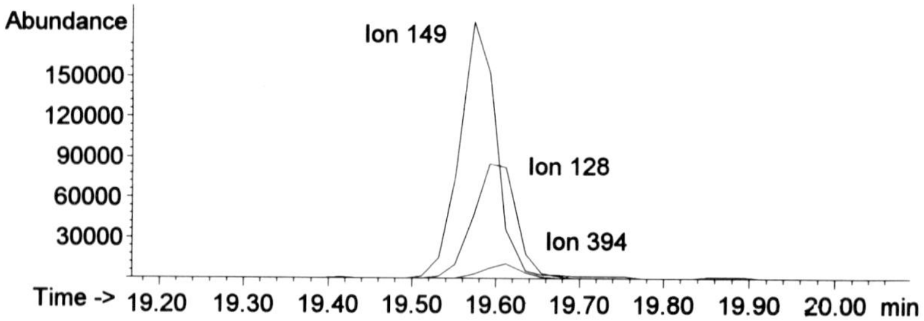

Fig 1. RIC of the "unknown" and the phthalate (Reproduced with permission from Ref. [49])

concentration level is also the reference for ground and even surface water. Therefore, a ground water sample from a depth of 2 m was spiked with 20 chlorinated pesticides and 16 triazine herbicides including a few other nitrogen-containing active compounds to give a concentration of 200 ng/l. After solid

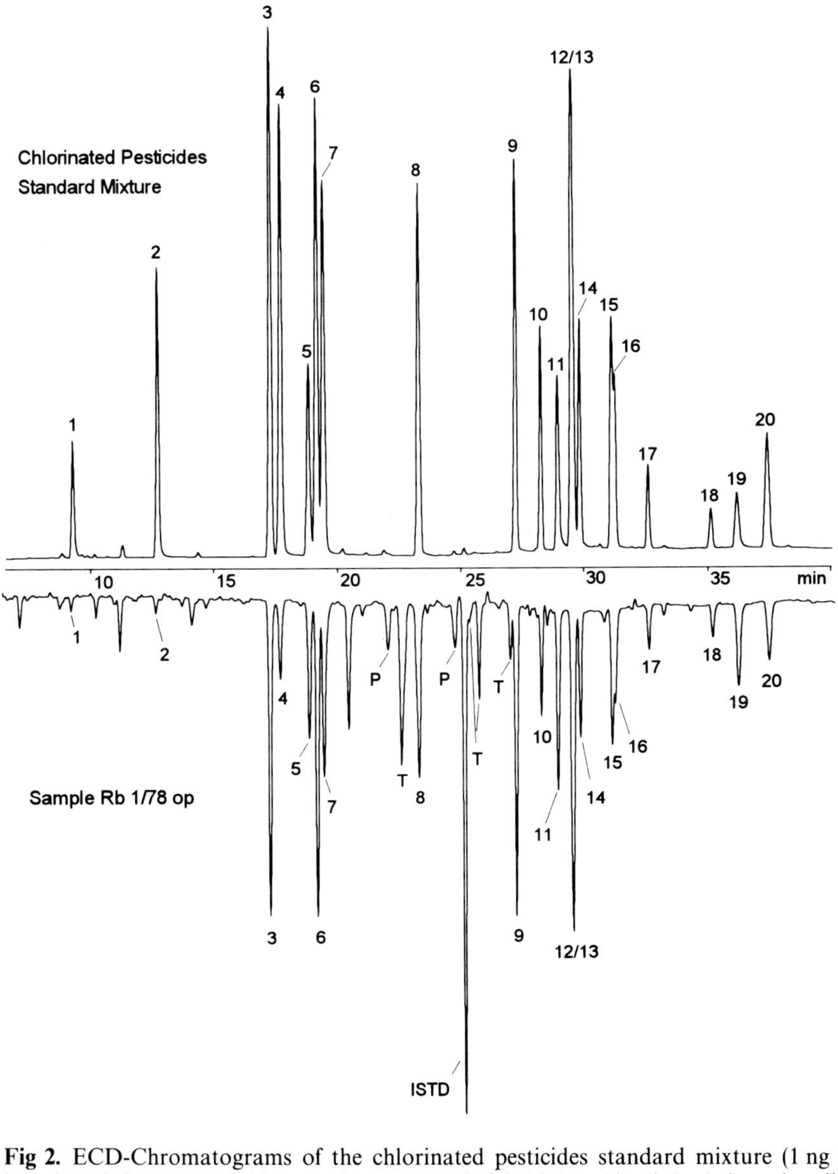

Fig 2. ECD-Chromatograms of the chlorinated pesticides standard mixture (1 ng injected) and the sewage field ground water sample Rb 1/78 op (depth 2 m). Sample spiked with 200 $ng\,l^{-1}$ (Chromatograms normalized) *T*: Peaks originating from the triazines mixture *P*: Peaks originating from phthalates

phase extraction (SPE) with RP-C18 adsorbent the extracts were concentrated to result in the injection of 1 ng of each analyte with the sample providing a full recovery. In Fig. 2 the chromatograms of a screening analysis on a 25 m fused silica column coated with SE 54 and connected to an ECD are shown.

The upper trace represents the standard mixture and the lower trace the extract from the ground water sample. All pesticides appear well separated with the exception of the two critical pairs dieldrin/p,p'-DDE (peaks 12, 13) and p,p'-DDD/o,p'-DDT (peaks 15 and 16). The chlorinated pesticides included in the standard mixture are listed in Table 2.

Note that all the analyses reported were performed on gas chromatographic systems continuously in use for multiresidue analysis of pesticides in water, soil and food samples. The chromatograms, therefore, do not demonstrate the highest separation efficiency achievable with fresh columns but give a realistic picture of results obtained in daily routine analysis.

Furthermore, the analytical method was not elaborated for the recovery of volatile chlorinated pesticides employing keeper addition because it is usually applied to triazine residue analysis in our laboratory. This can be easily depicted

Table 2. GC and GC-MS Data of chlorinated pesticides

No.:	Compound:	Rt(ECD) in min	Rt(MSD) in min	Mass 1 (abund.)*	Mass 2 (abund.)*	Mass 3 (abund.)*
1	Dichlobenil	9.26	8.63	171 (100)	173 (65)	136 (21)
2	Pentachlorobenzene	12.73	12.88	250 (100)	252 (63)	248 (63)
3	alpha-HCH	17.29	18.09	181 (100)	183 (98)	219 (91)
4	Hexachlorobenzene	17.71	18.63	284 (100)	286 (80)	282 (55)
5	beta-HCH	18.82	19.69	183 (99)	181 (95)	219 (80)
6	Lindane	19.14	20.04	183 (100)	181 (99)	219 (79)
7	Quintozene	19.43	20.45	237 (100)	295 (99)	297 (62)
8	Heptachlor	23.27	24.43	272 (60)	274 (47)	372 (4)
9	Heptachlorepoxid-trans	27.21	27.57	183 (96)	185 (76)	353 (19)
10	o,p-DDE	28.26	28.57	246 (100)	248 (66)	318 (30)
11	Chlorfenson	28.94	29.17	175 (99)	177 (36)	302 (17)
12	p,p-DDE	29.55**	29.60	246 (100)	248 (67)	318 (62)
13	Dieldrin	29.55**	29.78	79 (100)	263 (18)	277 (15)
14	o,p-DDD	29.83	29.92	235 (100)	237 (66)	165 (53)
15	p,p-DDD	31.13**	31.13	235 (100)	237 (68)	320 (2)
16	o,p-DDT	31.13**	31.27	235 (100)	237 (68)	354 (2)
17	p,p'-DDT	32.60	32.64	235 (100)	237 (65)	165 (46)
18	Methoxychlor	35.11	35.36	227 (100)	228 (17)	344 (1)
19	Tetradifon	36.16	36.72	229 (91)	356 (90)	354 (68)
20	Mirex	37.36	38.37	272 (100)	274 (77)	237 (52)
	Aldrin (ISTD)	25.03	26.00	66 (100)	263 (53)	265 (34)

* In brackets: relative abundancies of the ions in full mass spectra
** compounds coeluting

from the lower trace in Fig. 2 representing the analysis of the spiked ground water. Dichlobenil and pentachlorobenzene were found to be nearly totally lost and hexachlorobenze with low recovery. The other chlorinated pesticides were easily detected at this concentration level whereas triazine herbicides could not be analyzed by applying ECD. Metribuzine is the only compound of this group that is detectable with good response, the other pesticides that respond to the ECD are metolachlor, triadimefon and metazachlor. They are indicated in the chromatogram as well as the internal standard aldrin (ISTD) which is added to all samples before extraction to check the whole analytical procedure.

In Fig. 3, the chromatograms of the screening analysis with the same spiked ground water sample with NPD detection are reported.

In the lower trace the standard mixture of triazines and a few N-containing pesticides as listed in Table 3 and in the upper trace the chromatogram of the residue analysis are shown.

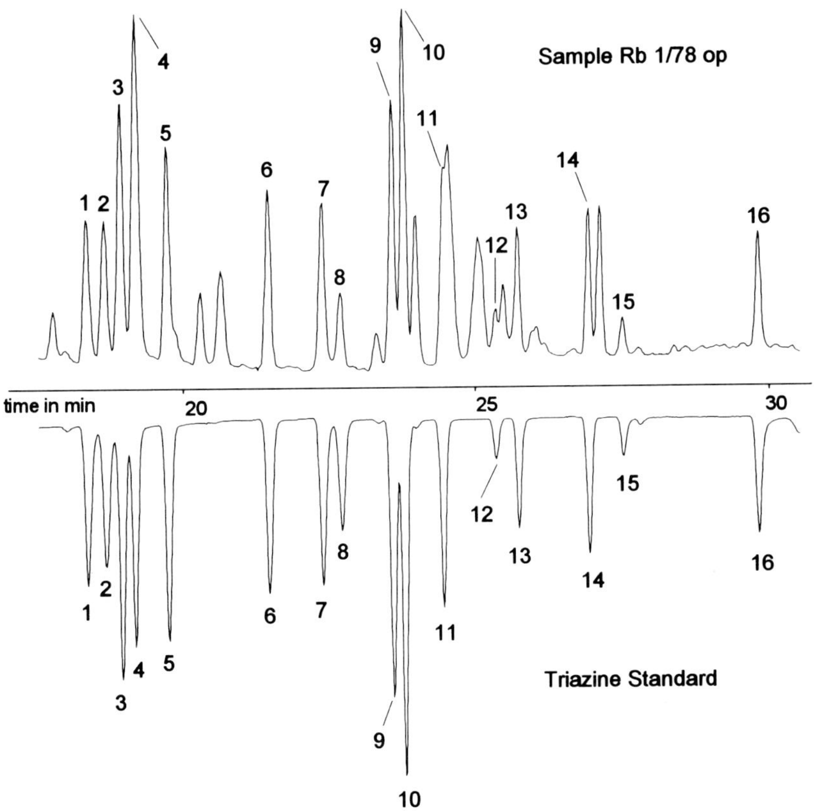

Fig 3. NPD-Chromatograms of the triazine standard mixture (1 ng injected) and the sewage field ground water sample Rb 1/78 op (depth 2 m). Sample been spiked with 200 ng l^{-1}

Table 3. GC and GC-MS Data of triazine herbicides

No.:	Compound:	Rt(NPD) in min	Rt(MSD) in min	Mass 1 (abund.)*	Mass 2 (abund.)*	Mass 3 (abund.)*
1	Atraton	17.84	18.70	196 (100)	211 (86)	169 (56)
2	Simazine	18.17	19.17	201 (100)	186 (57)	203 (33)
3	Atrazine	18.46	19.44	200 (100)	215 (53)	202 (32)
4	Propazine	18.69	19.71	214 (100)	229 (62)	216 (31)
5	Terbuthylazine	19.29	20.33	214 (100)	173 (45)	229 (28)
6	Sebuthylazine	21.06	22.25	200 (100)	202 (34)	229 (15)
7	Desmetryn	22.01	23.40	213 (100)	198 (63)	171 (45)
8	Metribuzine	22.36	23.62	198 (100)	199 (31)	214 (8)
9	Ametryn	23.27	24.47	227 (100)	212 (65)	170 (43)
10	Prometryn	23.47	24.63	241 (100)	184 (98)	226 (56)
11	Terbutryn	24.17	25.26	226 (100)	185 (50)	241 (44)
12	Metolachlor	25.12	26.11	162 (100)	238 (48)	240 (15)
13	Triadimefon	25.52	26.43	57 (100)	208 (26)	210 (9)
14	Metazachlor	26.77	27.51	209 (100)	133 (94)	277 (15)
15	Triadimenol	27.39	27.96	112 (100)	168 (58)	128 (45)
16	Methoprotryne	29.76	29.91	256 (100)	271 (20)	213 (32)

*In brackets: relative abundancies of the ions in full mass spectra

All the target compounds are well separated and sensitively detected. In the ground water sample chromatogram a few additional peaks from the matrix are observed one of them obviously overlapping terbutryn (peak 11). Summarizing these results, it can be concluded that screening for chlorinated pesticides and triazine herbicides can be performed easily and reliably for ground water applying high performance gas chromatography with the selective detectors ECD and NPD. This holds true in particular when using SPE with carefully selected RP-C18 adsorbent. The extracts contain less coextractants than those obtained with liquid/liquid extraction (LLE) using dichloromethane or ethylacetate according to our long experience. It finally must be added that the screening analysis reported was performed on one gas chromatograph applying eluent splitting to the two detectors NPD and ECD in parallel. The information from one detector trace is always complementary to that of the other. Note further that the same spiked water samples were analyzed with GC-AED and reported in chapter 9 to allow comparison and sound judgement of the various equipment available for screening analysis.

5.7 Confirmation with Full Scan

The extract of the spiked water sample was subjected to confirmatory analysis with GC-MS is full scan in order to try to identify the pesticides by automated library search. The total ion current (TIC) chromatogram obtained using a 25 m fused silica column (SE-54) connected to a mass selective detector (HP MSD) is shown in Fig. 4.

The chromatogram is dominated by two large peaks at 23, 85 and 26,20 min that originate from di-isobutyl and di-butyl phthalate contaminants ubiquitously found in residue analysis.

Fully automated confirmatory analysis can be carried out with modern instruments. Firstly, an automated integration procedure indicated a total of 73 peaks in the chromatogram. Then automated library search was performed with a target compound mass spectral library (HP Pesticide Library, [56]) and varying parameter setting and various background subtraction procedures. At the end a total of only 5 out of 33 pesticides easily detected in the screening analysis could be identified by their mass spectra of which two were dubious due to their poor quality. Propazine, promethryn and metazachlor were recognized with all the parameter settings optimized for trace level analysis. This is in accordance with our experience that full scan confirmation is not feasible at the low concentration level as set by the EC drinking water guideline. The merits of full scan application are in the identification of unknown contaminants revealed with the above screening procedure and which are frequently referred to as unidentified analytical responses (UAR) or in the confirmation of pesticides or environmental contaminants observed as larger peaks in the TIC chromatogram. An example will be given later in this chapter showing the discovery of a new contaminant in ground and surface water.

5.8 Confirmation with Selected Ion Monitoring (SIM)

SIM is a measurement method which makes the mass spectrometer into a highly selective detector tuned to monitoring a small number of mass channels. The

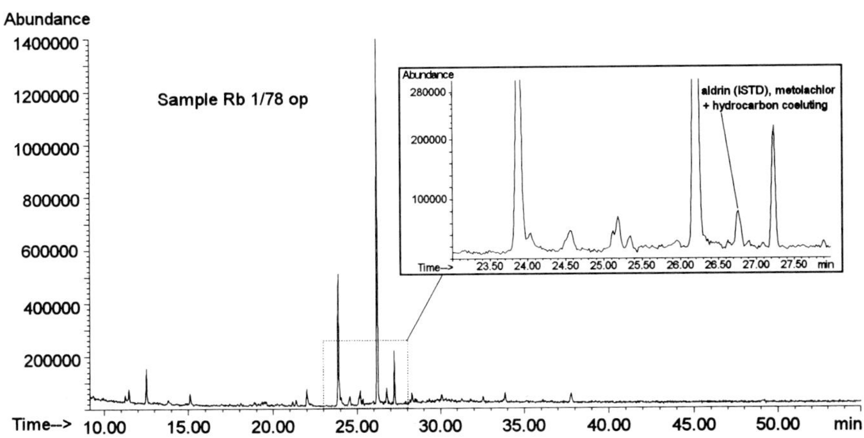

Fig 4. TIC chromatogram of the sewage field ground water sample spiked with 200 ng l^{-1} of the standard mixtures as in Figs. 2 and 3

high gain in detection sensitivity more than compensates for the reduction of structural information. This has proven again with the easy SIM confirmation of all pesticides added to the ground water of which three examples are given here. In Fig. 5, the confirmation of p,p'-DDE by means of the 3 indicative ions m/z 318, 248 and 246 monitored in a time window between 29.35 and 30.50 min is shown and in Fig. 6 that of dieldrin monitored in the same time window but with ions m/z 380, 263 and 79.

Note that with the gaschromatographic column used in the confirmatory analysis the critical pair is well resolved in contrast to the screening analysis presented in Fig. 2. This fact is, however, of no importance in the confirmatory analysis because of the specificity introduced with the mass selective detection which makes the detection of p,p'-DDE transparent for dieldrin and vice versa that of dieldrin transparent for p,p'-DDE. The reduction of structural information caused by measuring only three ions instead of whole spectra is sometimes overestimated because mass spectral identification is based on various criteria.

Reproducibility of retention times of better than 0.1 min is easily achieved with capillary columns that provide the highest chromatographic separation power of all chromatographic techniques, and can be additionally checked with internal standards. All three ions must give rise to a peak at the retention time corresponding within the reproducibility margin with the reference compound. Furthermore, the appearance profiles should be uniform and clearly resolved from other sample compounds where they have ions in common. The three

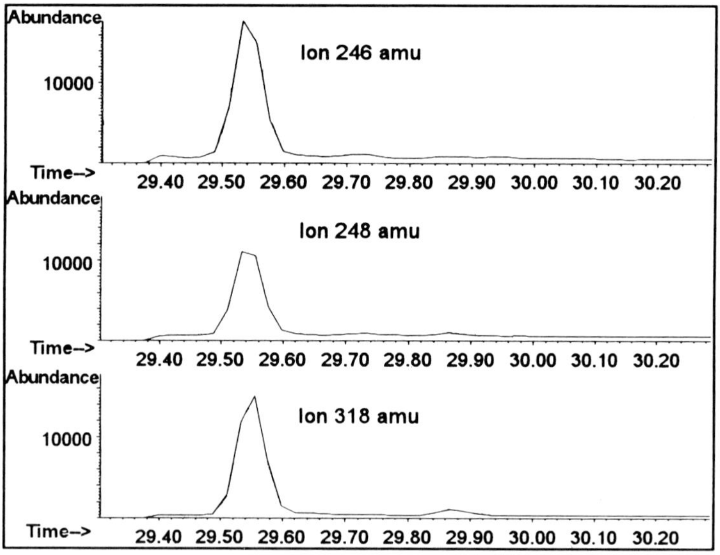

Fig 5. Ion traces (m/z 246, 248, 318) of p,p'-DDE recorded with SIM. Sewage field ground water sample spiked as in Fig. 2

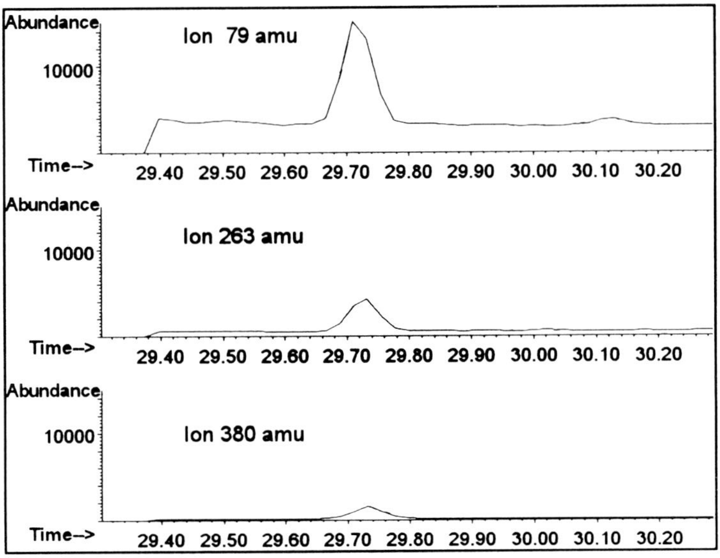

Fig 6. Ion traces (m/z 79, 263, 380) of dieldrin recorded with SIM. Sewage field ground water sample spiked as in Fig. 2

selected ions must also match in relative intensity. Correspondence at trace level concentrations is considered as established if the maximum difference intensity ratios of indicative ions relative to the reference is less than 20% [57]. The relative ratios can be determined either as peak heights or as areas of the ion chromatogram peaks. The use of isotope peaks of chlorine for this comparison is also acceptable for confirmation of chlorinated pesticides as demonstrated with ions m/z 246 and 248 in the SIM analysis of p,p'-DDE.

Ions selected for SIM confirmation must be intense in the mass spectrum and indicative, that means all ions prominent with the mass spectrometric background from column bleed and common environmental contaminants as phthalates and hydrocarbons should be avoided. A compilation of such background ions is given in Table 4.

In detail, possible background interferences are referred to by Budzikiewicz [58] who presents mass spectra of solvents, various column coatings, silicon oil and a hydrocarbon mixture. Generally, ions with higher masses are to be preferred because of their statistically lower abundance in other compounds and consequently greater significance. Ions of lower mass can arise by fragmentation from many compounds with higher molecular weight. A molecular ion of sufficient high intensity is mostly the best suited indicative ion as applied in the confirmation of p,p'-DDE with m/z 318. The confirmation of dieldrin, however, relies also on an ion representing a fragment of rather low mass, namely m/z 79 which constitutes the base peak in the spectrum. This mass is not found within

Table 4. List of ions originating from common background interferences

Ions (m/z)	Possible source of interferences
149, 167, 279	plastizisers (phthalic esters)
129, 185, 259, 329	plastizisers (tributyl acetyl citrate)
99, 155, 211	plastizisers (tributyl phosphate)
73, 133, 147, 207, 221, 281, 355, 429, 503	silicon oil
	column coating (SE-30, SE-54, OV-101, OV-1)
135, 197, 209, 259, 333, 345, 408, 465, 527	column coating (OV-17, OV-11)
75, 91, 135, 156, 169, 183, 253, 352,389	column coating (OV-225)
449, 458, 502, 511, 520	
29, 43, 57, ...;41, 55, 69, ...;53, 67, 81,...	hydrocarbons
39, 45, 51, 65, 91, 92	solvent (toluene)
64, 96, 128, 160, 192, 224, 256	sulfur

the common fragment ion series originating from alkanes or aromatic compounds and is therefore acceptable. The proof of suitability, however, is always the appearance of the ion trace in the appropriate time window. As can be seen from figure 6, the background noise in the ion trace of m/z 79 is considerably higher than those of the two higher masses but more important is a good signal to noise ratio and that no other peak is observed in the time window selected for dieldrin confirmation.

5.9 Target Compound Analysis with SIM

The combination of the high separation efficiency provided by modern capillary columns with tuneable selectivity and high sensitivity provided by mass spectrometric detectors under SIM conditions has gained reputation as the most powerful tool in ultra trace analysis in recent years. It is the analytical method that allows the surveillance of water with respect to the low maximum tolerances established for pesticides in drinking water. The method takes advantage from of the SIM time window programming provided with modern computerized instruments and the high reproducibility of retention times.

5.9.1 Chlorinated Pesticides

To demonstrate the merits of target pesticide analysis with SIM and time window programming, the same spiked ground water sample was analyzed as in the examples described with the screening and confirmatory analysis. The SIM program applied for the analysis of 21 chlorinated pesticides is compiled in table 5.

The resulting chromatograms are presented in Fig. 7.

Table 5. SIM programming parameters and retention times for chlorinated pesticide analysis

Time Window	Compound	t_R [min]	Time Window [min]	Selected ions per window [m/z]	Dwell time per ion [ms]
1	Dichlobenil	8.63	8.50–11.99	136, 171, 173	200
2	Pentachlorobenzene	12.88	12.00–15.99	248, 250, 252,	200
3	alpha-HCH	18.09	16.00–19.19	181, 183, 219,	150
	Hexachlorobenzene	18.63		228, 284, 286	
4	beta-HCH	19.69	19.20–22.99	181, 183, 219	150
	Lindane	20.04		237, 295, 297	
	Quintozene	20.45			
6	Heptachlor	24.43	23.00–25.49	272, 274, 372	200
7	Aldrin (ISTD)	26.00	25.50–27.99	66, 183, 185,	150
	Heptachlorepoxid-trans	27.57		263, 265, 353	
8	o,p-DDE	28.57	28.00–28.89	246, 248, 318	200
9	Chlorfenson	29.17	29.00–29.36	175, 177, 302	200
10	p,p-DDE	29.60	29.37–30.49	79, 235, 237,	100
	Dieldrin	29.78		246, 248, 263,	
	o,p-DDD	29.92		277, 318	
11	p,p-DDD	31.13	30.50–33.99	165, 235, 237,	170
	o,p-DDT	31.27		320, 354	
	p,p'-DDT	32.64			
12	Methoxychlor	35.36	34.00–35.89	227, 228, 344	200
13	Tetradifon	36.72	35.90–37.49	229, 354, 356	200
14	Mirex	38.37	37.50–40.00	237, 272, 274	200

The chromatograms do not look like the typical SIM presentations as shown in Fig. 4 and 5 but like common TIC chromatograms. They result in fact from the summation of all detected ions, a technique referred to as multiple ion detection (MID) in order to differentiate from TIC which is the analog with full scan mode. The appearance of the MID chromatograms depend very much on the ions selected for the SIM analysis whereas that of the TIC chromatograms reflects the burden of co-extractives in an environmental sample. Although the MID chromatograms appear almost transparent to coextracting matrix compounds, it is a severe mistake to apply an MID peak for quantitation of an analyte without having carefully checked the peak shapes and peak area ratios of all indicative ions. Quantitation must be performed with the ion least interfered with by matrix compounds or with all three ions independently. The latter method provides at the same time a good indication of interferences should one ion trace give a different result.

In Fig. 7 the upper trace shows the MID chromatogram of the standard mixture of 18 chlorinated pesticides compared to the extract from the spiked ground water sample in the lower trace. The two chromatograms appear very similar with the pesticides exhibiting as major peaks in the water extract. This is a picture totally different from that shown in the full scan chromatogram as

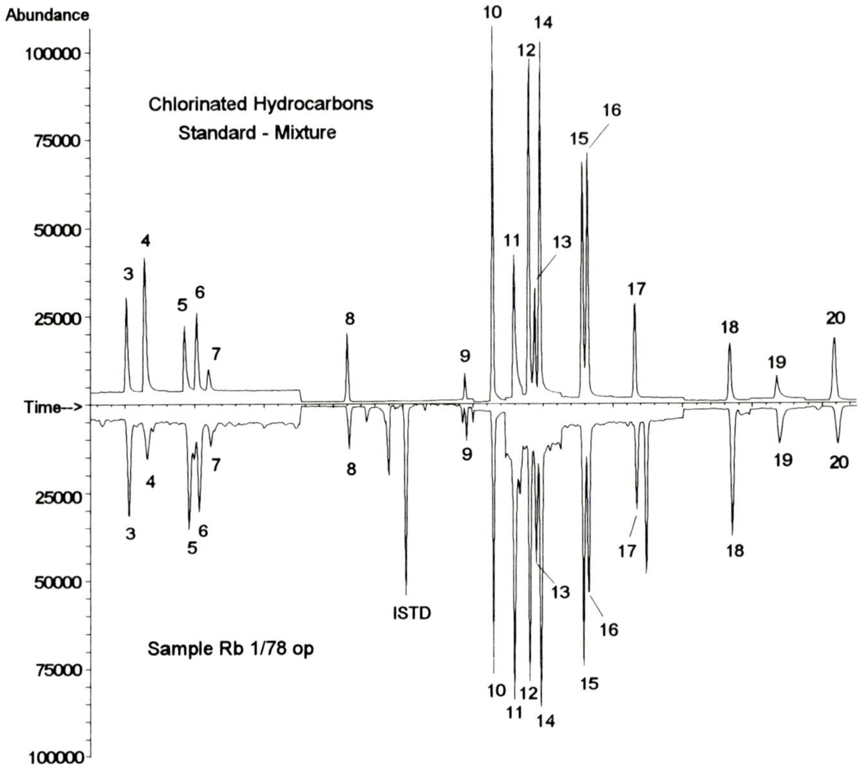

Fig 7. MID chromatograms recorded with SIM of a chlorinated pesticide standard (2 ng injected) and the sewage field ground water sample, spiked with standard mixtures at a concentration level of 200 $ng\,l^{-1}$ as in Fig. 2

presented in figure 4. There the pesticide peaks disappear completely in the common background. Note the different baseline heights in the individual time windows which are, as expected, higher in the water sample chromatogram. Only a few additional peaks are observed in the sample chromatogram indicating the high degree of transparency for the co-extracts from the matrix. This makes the MID chromatograms of figure 7 similar to the ECD chromatograms shown in Fig. 2. The relative sizes of the peaks in the chromatograms, however, are not comparable because of the completely different mechanism of selective detection. Quintozene for instance constitutes one of the major peaks in the ECD chromatograms but only a small one in the MID chromatogram. The reason for the small size of the peak in the MID chromatogram is the high extent of fragmentation spreading the total ion intensity to many fragments of which only three can be selected for SIM. They represent a corresponding low proportion of the TIC. This illustrates a general rule, compounds with moderate fragmentation exhibiting the molecular ion and high mass fragments of high intensity in their mass spectra can be detected with lowest detection limits with SIM.

5.9.2 Triazine Herbicides

In Fig. 8, the MID chromatograms of the triazine herbicide standard mixture and the corresponding extract from the spiked ground water is presented.

The SIM program applied for the analysis of 16 triazine and other N-containing pesticides is presented in Table 6.

In both chromatograms the major peaks are those representing the pesticides. The co-extracts from the matrix remain completely transparent, comparable to the appearance of the NPD chromatograms shown in Fig. 3. However, there is an important difference between the two selective chromatograms. A peak in the NPD chromatogram indicates a nitrogen or phosphorus containing compound or even a co-extract breaking the cross-selectivity boundary (see chap 3), more information about the compound's nature is not provided. In the peak of an MID chromatogram the full information of SIM is available as used for confirmation analysis. Each peak can easily be resolved into the three indicative ions selected for SIM analysis as shown in Fig. 9 for peak 11 from the MID chromatogram of the ground water extract in Fig. 8.

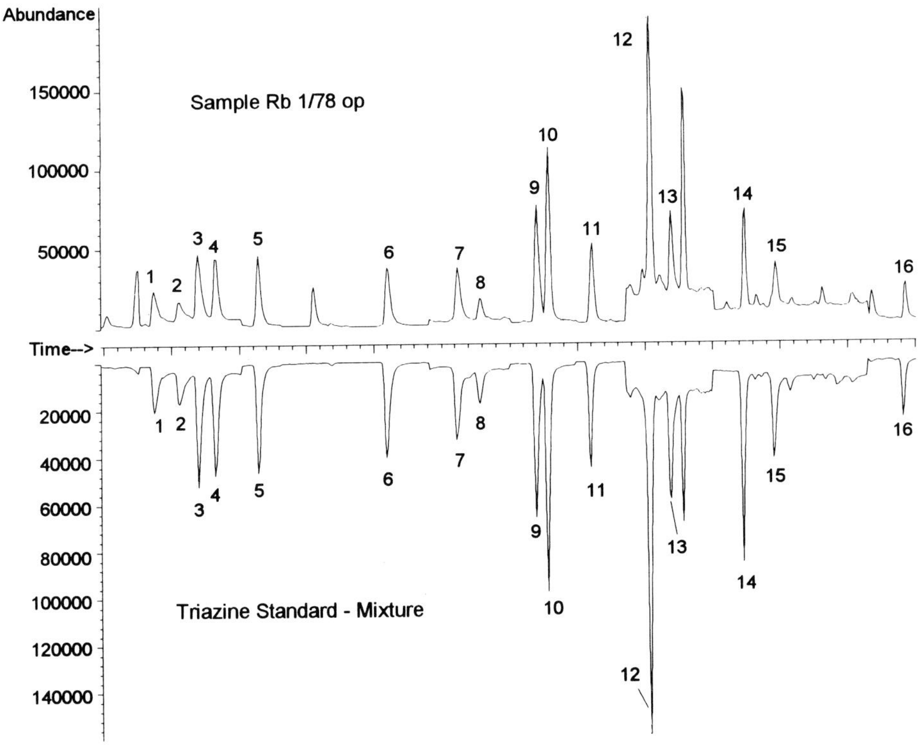

Fig 8. MID chromatograms recorded with SIM of a triazine standard mixture (2 ng injected) and the sewage field ground water sample, spiked with standard mixtures at a concentration level of 200 ng l^{-1} as in Fig. 3

Table 6. SIM programming parameters and retention times for triazine herbicide

Time Window	Compound	t_R [min]	Time Window [min]	Selected ions per window [m/z]	Dwell time per ion [ms]
1	Atraton	18.70	18.50–18.84	169, 196, 211	200
2	Simazine	19.17	18.85–19.99	186, 200, 201,	100
	Atrazine	19.44		202, 203, 214,	
	Propazine	19.71		215, 216, 229	
3	Terbuthylazine	20.33	20.00–20.99	173, 214, 229	100
4	Sebuthylazine	22.25	21.00–22.79	200, 202, 229	
5	Desmetryn	23.40	22.80–23.99	171, 198, 199,	150
	Metribuzine	23.62		213, 214	
6	Ametryn	24.47	24.00–24.94	170, 184, 212,	
	Prometryn	24.63		226, 227, 241	
7	Terbutryn	25.26	24.95–25.69	185, 226, 241	200
8	Metolachlor	26.11	25.70–26.99	57, 162, 208,	150
	Triadimefon	26.43		210, 238, 240	
9	Metazachlor	27.51	27.00–29.29	112, 128, 133,	150
	Triadimenol	27.96		168, 209, 277	
	Methoprotryne	29.91	29.30–30.50	213, 256, 271	200

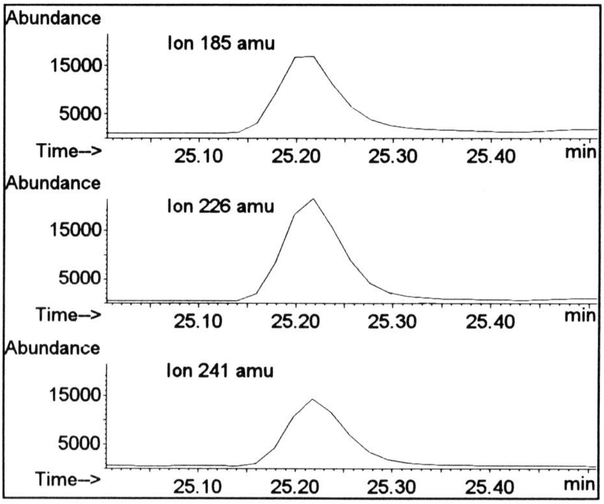

Fig 9. Ion traces (m/z 185, 226, 241) of terbutryn recorded with SIM (peak 11 in the MID chromatogram of the spiked sewage field ground water sample in Fig. 8)

The appearance of the 3 ion peaks at the calibrated retention time and their fitting intensity ratio prove the presence of the target compound terbutryn. In the same way, all the other 34 pesticides were easily confirmed in the ground water extract. The few co-extractives appearing in the MID chromatograms in

Fig. 7 and 8 could easily be identified as not being target pesticides because a peak usually only appears in one ion trace. Therefore, we consider the presentation of SIM results in MID chromatograms to be a most adequate form. These clear chromatograms exhibit usually only a few peaks that must be checked with the 3 corresponding ion traces as can be depicted from MID chromatograms of examples reported later in this chapter (Fig. 14 and 15).

5.9.3 Phenoxyalkanoic Acids

Phenoxyalkanoic acids and other acidic herbicides are of particular interest in water analysis because of their widespread use in agriculture and urban weed control. Screening analysis of this group of pesticides is possible with GC methods after derivatization. Problems arose from the fact that with the easily obtainable methyl esters there is a lack of detection sensitivity experienced for some major phenoxyalkanoic acids (in particular monochlorinated compounds) when applying ECD detection. Other selective detectors are not available, with the exception of the recently introduced AED. Pentafluorobenyzl (PFB) esters have gained importance in overcoming the sensitivity problem with the ECD. However, especially in water analysis, a large number of interfering substances are formed from the environmental matrix when the extract is to be analyzed with ECD. One new approach is the determination of the PFB esters with AED element specific detection of chlorine and other heteroatoms usually not found in the environmental extracts as described in chapter 8.

Another approach is target compound analysis applying SIM which enabled us to achieve detection levels in the low picogram region [51]. The method was found to be applicable to all kinds of water samples when using an optimized SPE with RP-C 18 adsorbent [52].

In Fig. 10, the MID chromatograms of the standard mixture of 18 PFB esters (lower trace) and of the extract of a spiked ground water sampled from a sewage field with additional pentafluorobenzylation are shown.

Both chromatograms are dominated by the pesticide peaks with only a few minor contaminant peaks in the water sample. The acidic compounds analyzed as their PFB esters are compiled in Table 7, they include 15 herbicides, two internal standards and the pharmaceutical clofibric acid which has recently been found in many water samples.

The discovery of this widespread water contaminant will be described in the next section.

Following the first finding of clofibric acid in ground water samples in Berlin it was necessary to develop a very sensitive detection method for this target compound in order to investigate the distribution of this compound in the aqueous system. In Table 7, all the indicative ions for the target compounds that can be applied in the SIM method are listed. The base peak in the mass spectra of most PFB esters is formed with the pentafluorobenzylium ion m/z 181 which exhibits great stability by favourable mesomeric resonance (see chap 5).

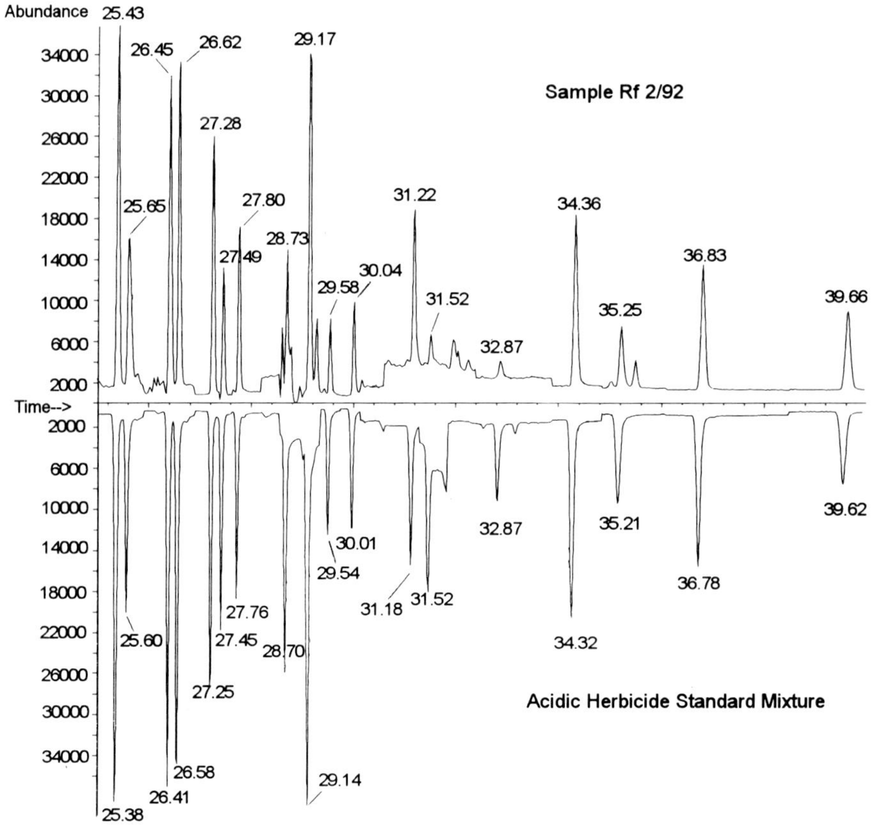

Fig 10. MID chromatograms recorded with SIM of the PFB-esters of an acidic herbicide mixture (2 ng injected) and a sewage field ground water sample spiked with 200 ng l^{-1}

The ion originates also from many interfering substances which during derivatization form also PFB esters. Therefore, the ion m/z 181 was not used as an indicative ion in the SIM method. A thorough investigation of water samples of different origin resulted in better signal to noise ratio of the base line in the MID chromatogram and almost full transparency to environmental coextractives [53]. This can be easily depicted from a derivatization blank as demonstrated in Fig. 11.

While the MID chromatogram of the time programmed SIM method shows only the characteristic steps in a flat baseline, the appearance of the ion trace at m/z 181 is that of a common gas chromatogram of a mixture of compounds. The peaks seen represent impurities from the reagents and by-products formed during derivatization. Many more peaks appear in water samples. In ultra trace analysis the feature of an adjustable selectivity as here demonstrated with SIM is absolutely necessary and not provided by a non-selective detector as an ECD.

Table 7. Mass spectrometric data of the PFB derivatives

No.:	Compound:	MW	m/z 181 (abund.)*	Mass 1 (abund.)*	Mass 2 (abund.)*	Mass 3 (abund.)*
1	Clofibric acid	394	38	128 (100)	130 (32)	394 (13)
2	2,4-Dichlorobenzoic acid	370	100	173 (48)	370 (42)	372 (28)
3	Mecoprop	394	82	169 (100)	394 (94)	142 (59)
4	Surrogate**	394	100	128 (90)	169 (66)	394 (77)
5	Dicamba	400	100	203 (79)	400 (31)	402 (20)
6	MCPA	380	100	380 (95)	141 (83)	382 (32)
7	Dichlorprop	414	100	162 (65)	414 (46)	416 (32)
8	2,4-D	400	100	175 (48)	400 (42)	402 (27)
9	1-Naphthylacetic acid	366	17	141 (100)	366 (84)	115 (19)
10	Triclopyr	435	100	210 (30)	435 (20)	437 (19)
11	Fenoprop	450	100	196 (46)	448 (24)	450 (24)
12	Bentazon	420	100	378 (33)	420 (15)	341 (11)
13	MCPB	408	100	267 (25)	142 (5)	408 (3)
14	2,4-DBD	428	100	267 (24)	162 (8)	428 (1)
15	Fluazifop	507	21	282 (100)	254 (58)	507 (53)
16	Benazolin	423	45	170 (100)	423 (87)	198 (2)
17	Haloxyfop	541	30	316 (100)	288 (63)	541 (44)
18	Flamprop	501	6	105 (100)	276 (11)	501 (4)

* relative intensity of the charged mass fragment in the EI spectrum, basepeak = 100
** 2-(4-Chlorophenoxy) butyric acid

All the problems which confront the analyst when using GC-ECD to analyze PFB esters arise from the contaminants that can be addressed with ion trace m/z 181. The SIM programming parameters used in the target analysis of acidic compounds are compiled in Table 8.

The method has been successfully applied for an extensive screening of acidic herbicides and clofibric acid in surface, ground and tap water achieving an instrumental detection limit of 1 to 10 pg for all compounds. With the SPE procedure presented in Fig. 12, the limits of detection in water samples were determined to be between 1 ng and 10 ng per liter.

In Fig. 13, the MID chromatogram of a Berlin tap water sample containing 20 ng l^{-1} clofibric acid is presented.

The MID chromatogram demonstrates not only the reported detection sensitivity but also the absence of any interfering contaminants in the entire chromatogram as stated.

The merits of the target compound analysis approach as described are demonstrated with the analysis of a water sample from the river Havel [53]. In a TIC chromatogram of a GC run with full scan data acquisition, as shown in Fig. 14, a great number of peaks are observed swamping all the target compounds.

The positions of clofibric acid (4-CP) and the standards added are indicated in the chromatogram. The same sample was again analyzed on the same GC-

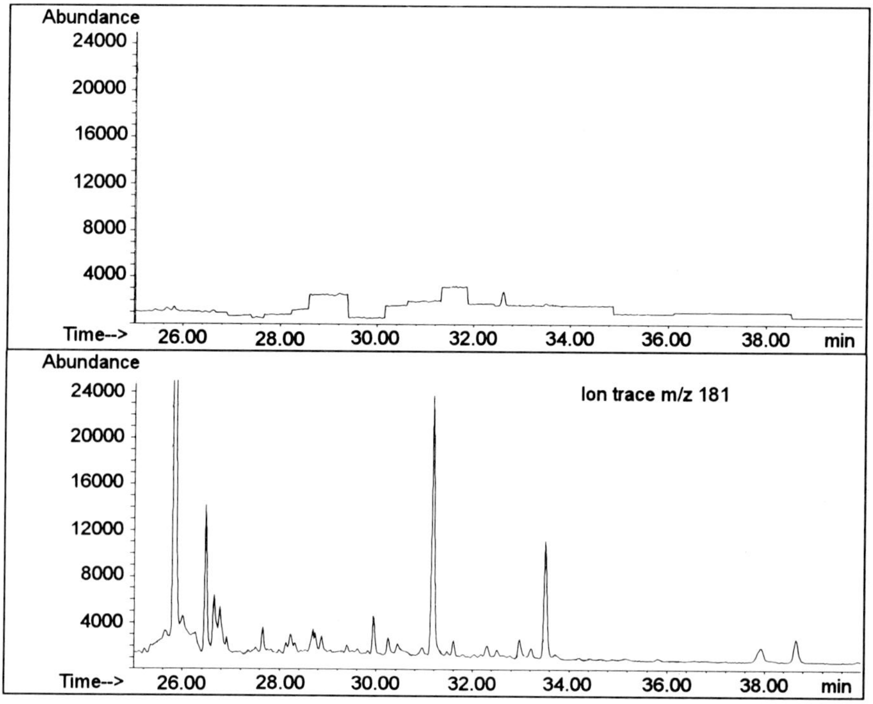

Fig 11. Derivatization blank of the pentafluorobenzylation *Top*: MID chromatogram of the indicative ions *Bottom*: Ion trace of m/z 181

MSD instrument applying the SIM method. The MID chromatogram shown in Fig. 15 looks totally different.

Three major peaks dominate the chromatogram representing the added internal and surrogate standard as well as clofibric acid originating from the river water at a concentration of 150 $ng\,l^{-1}$. The confirmation of the identity of all three compounds was easily performed by observing the indicative ion traces in the corresponding time window as shown in Fig. 16.

5.9.4 Elucidation of Unknown Contaminants by GC-MS

It often happens in screening analysis with selected detection that peaks of unknown identity are observed in the chromatograms. Such unknown analytical responses (UAR) are sometimes so conspicuous that their structural identity seems highly desirable. How the elucidation of the chemical structure of an unknown contaminant by means of GC-MS can be successfully achieved is best learned from a case history.

In the course of a monitoring of ground water samples for phenoxyalkanoic acid herbicides which were in widespread use at the former border areas in and

Table 8. SIM programming parameters and retention times for the acidic herbicide PFB-esters

Time Window	Compound	t_R [min]	Time Window [min]	Selected ions per window [m/z]	Dwell time per ion [ms]
1	Clofibric acid	25.42	25 –25.90	128, 130, 173,	150
	2,4-Dichlorobenzoic acid	25.65		370, 372, 394	
2	Mecoprop	26.43	25.91–26.89	142, 169, 394	200
	Surrogate	26.60			
3	Dicamba	27.28	26.90–28.49	141, 162, 203,	100
	MCPA	27.48		380, 382, 400, 402,	
	Dichlorprop	27.79		414, 416	
4	2,4-D	28.72	28.51–29.35	115, 141, 175,	150
	1-Napththylacetic acid	29.17		366, 400, 402,	
5	Triclopyr	29.56	29.36–29.74	210, 435, 437	200
6	Fenoprop	30.03	29.75–30.59	196, 448, 450	200
7	Bentazon	31.17	30.60–31.30	341, 378, 420,	200
8	MCPB	31.54	31.31–31.79	142, 267, 408	200
9	2,4-DB	32.87	31.80–33.89	162, 267, 428	200
10	Fluazifop	34.34	33.90–34.86	254, 282, 507	200
11	Benazolin	35.22	34.87–36.09	170, 198, 423	200
12	Haloxyfop	36.85	36.10–38.49	288, 316, 541	200
13	Flamprop	39.64	38.50–40.99	105, 276, 501	200

around Berlin an unknown chlorinated compound was found when applying GC-AED. Screening was carried out with the chlorine emission line in order to overcome the problems–already addressed in the former section–of the lack of specificity with the ECD. The application of GC-AED is described in chapter 7 and chapter 8. In the latter, the complementary part of the story is reported.

2,4-Dichlorobenzoic acid (2,4-DCB) was added to the water sample as a surrogate standard. The unknown chlorine-containing compound appeared in the chlorine trace chromatograms of many samples as a peak immediately before the 2,4-DCB-PFB ester. It was obviously an acid because it did not occur in the extract of neutral compounds and formed a PFB ester as could be depicted from the fluorine trace with the AED. Confirmatory analysis with GC-MS in full scan mode with automated library search gave no positive result but showed that the ubiquitous plasticizer di-isobutylphthalate eluted immediately in front of 2,4-DCB as shown in Fig. 17 with the TIC chromatogram of that derivatized sample extract.

Inspection of the individual scans over the whole peak gave the result of a chlorine containing compound overlapped by the phthalate. The chlorine containing fragment was easily recognized at m/z 128 with the isotope ion at m/z 130 as can be drawn from Fig. 18.

By means of reconstructed ion chromatograms (RIC), the appearance of all the ions observed in the mixed mass spectrum of di-isobutylphthalate (m/z 149

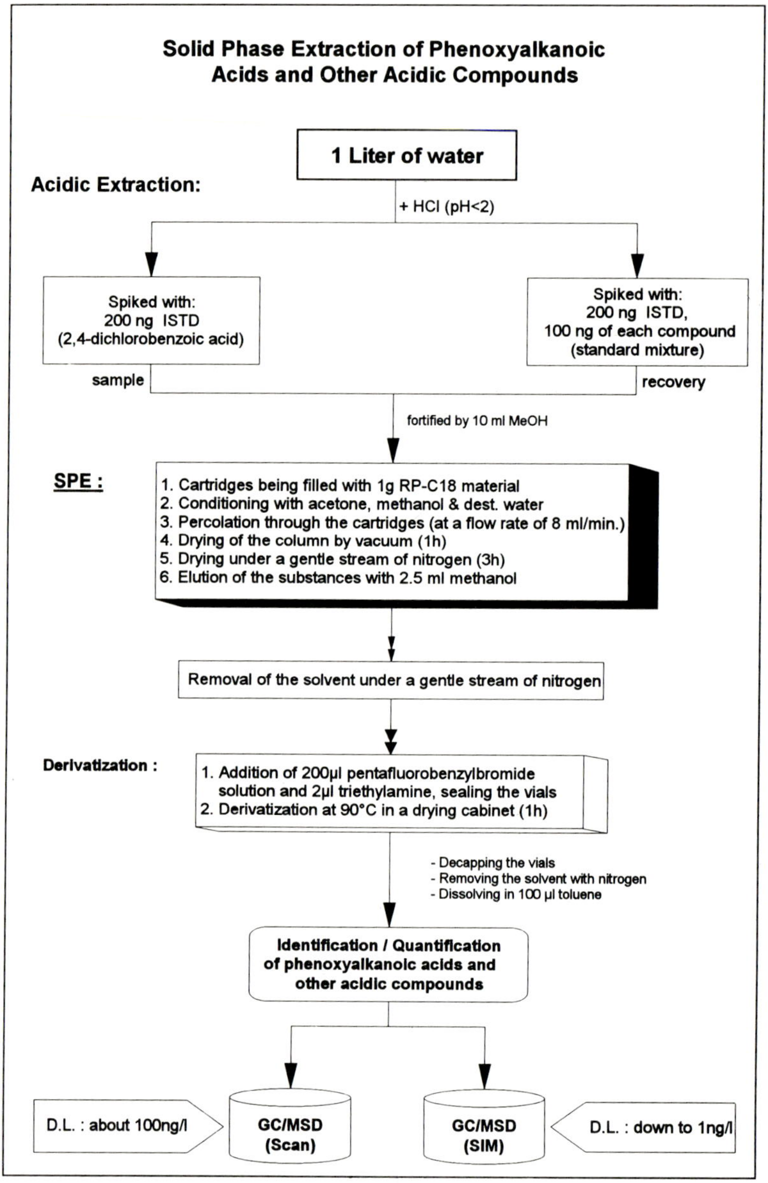

Fig 12. Solid phase extraction of acidic contaminants followed by pentafluorobenzylation and GC-MSD determination

and 278) and the unknown were inspected. It was found that the critical pair was somewhat resolved with all ions originating from the unknown showing the same peak shape and the same apex at 17.61 min while those of the phthalate eluted earlier. Note the important contribution of high performance GC separation in the recognition and elucidation of unknown contaminants.

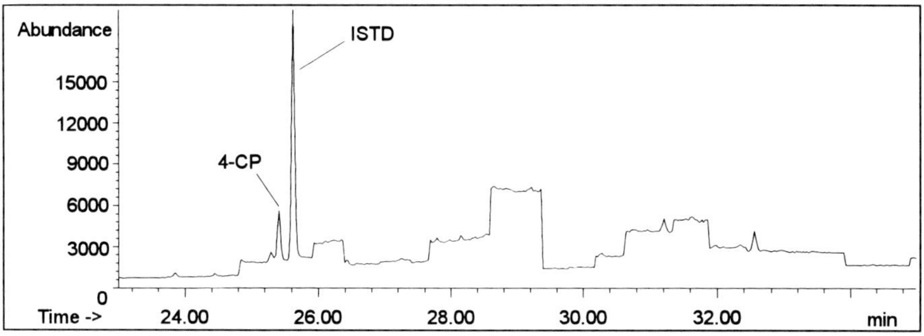

Fig 13. MID Chromatogram of a tap water sample containing 20 ng l^{-1} clofibric acid (4-CP). 200 ng l^{-1} 2,4 dichlorobenzoic acid added (ISTD) (Reproduced with permission from Ref. [53])

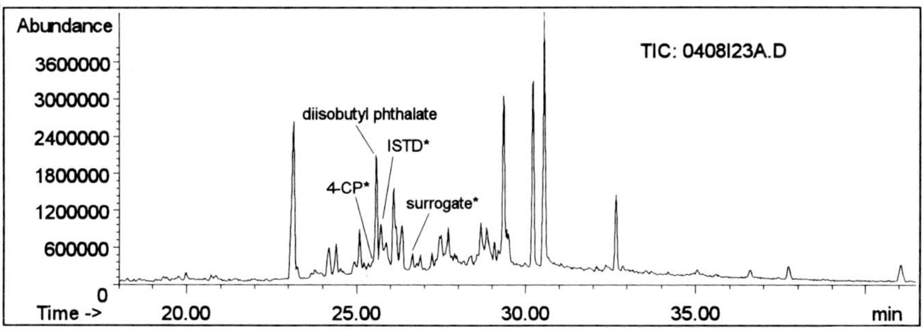

Fig 14. TIC chromatogram of a sample taken from the river Havel recorded in full scan mode (m/z 50-550). Sample spiked with 200 ng surrogate standard and 400 ng ISTD per liter. *Peaks resulting from matrix compounds overlapping the indicated target and standard compounds (Reproduced with permission from Ref. [53])

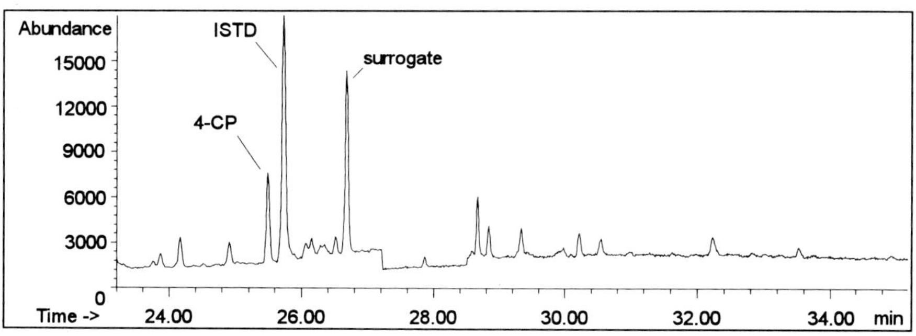

Fig 15. MID chromatogram recorded in SIM of the river water sample from Fig. 14. Sample with 74% recovery for the surrogate standard, containing 150 ng clofibric acid per liter (Reproduced with permission from Ref. [53])

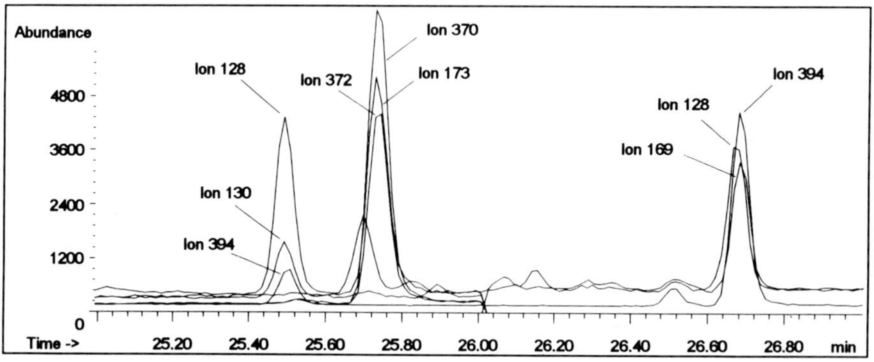

Fig 16. Ion traces taken from the MID chromatogram of the river water sample shown in Fig. 14 (Reproduced with permission from Ref. [53])

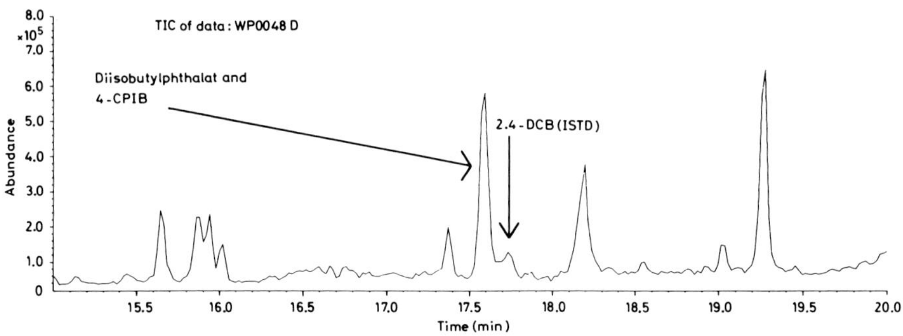

Fig 17. TIC chromatogram of a ground water sample recorded in full scan. One litre of the water sample was spiked with 2,4-dichlorobenzoic acid (internal standard), extracted and derivatized with pentafluorobenzyl bromide

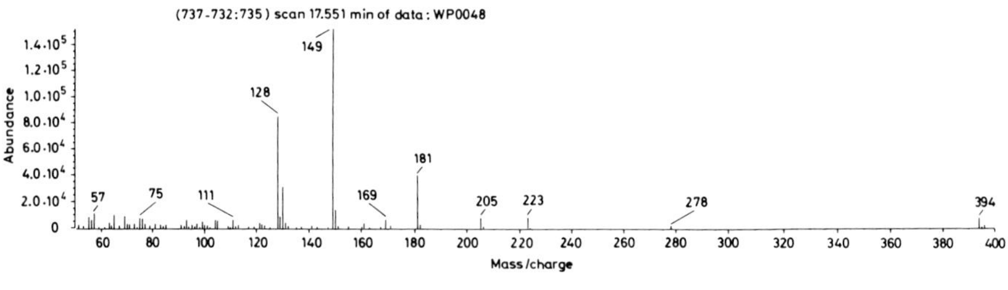

Fig 18. Spectrum of the peak at 17.55 min taken from the chromatogram in Fig. 17

The RIC traces of four ions indicative for the unknown are presented in Fig. 19. Consequently it was possible to generate a corrected mass spectrum from the full scan data showing the indicative ions of the unknown PFB-ester with only minor contribution from the phthalate (m/z 149) as shown in Fig. 20.

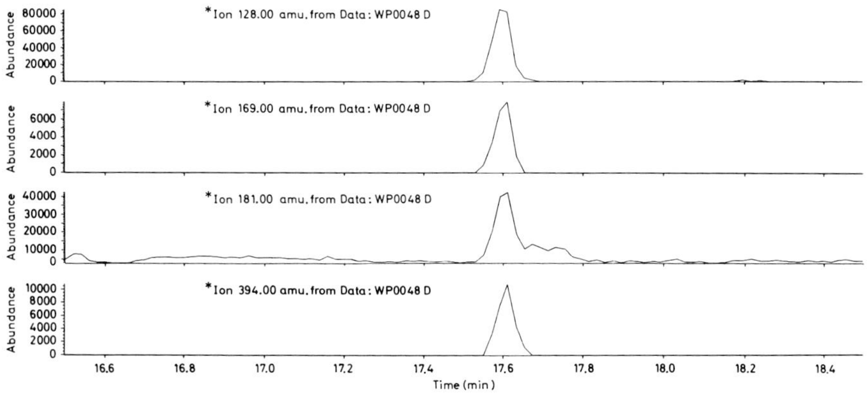

Fig 19. RICs of the four indicative ions of the unknown compound taken from Fig. 17

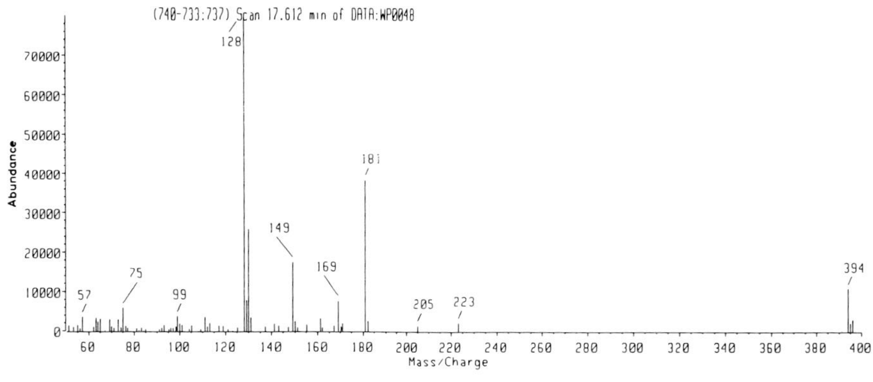

Fig 20. Background corrected spectrum of the unknown compound at 17.61 min taken from Fig. 17

The mass spectrum did not match any of those documented in our dedicated pesticide library. It resembled most the PFB ester of the herbicide mecoprop with the same molecular weight of 394 but the fragmentation pattern and the retention time was different. The interpretation of the mass spectrum resulted in the three main fragments as presented in Fig. 21.

The assumption of having the chlorine in the *para*-position corresponds to the knowledge that 4-chlorophenoxyalkanoic acids are potential herbicides. The structure of fragment ion m/z 169 was not clear; therefore, all isomeric 4-chlorophenoxy butyric acids were synthesized and their PFB esters measured. The PFB ester of 2-(4-Chlorophenoxy)-2-methyl propionic acid (4-CP) was found to match exactly the mass spectrum of the unknown and to fit best with its retention time. Finally, the isomers with the chlorine in position 2 and 3 at the

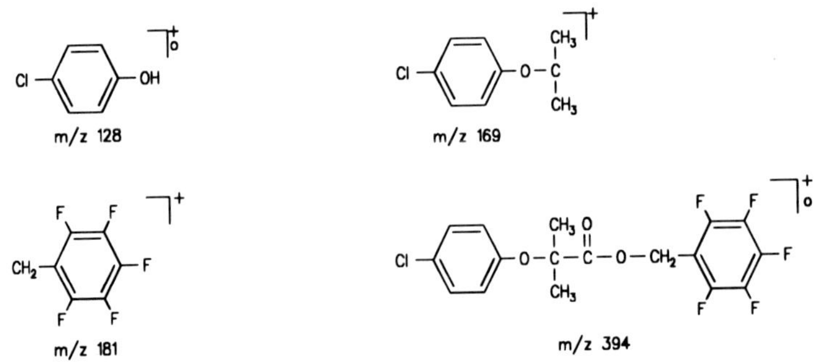

Fig 21. Main fragments and molecular ion of the former unknown compound

phenoxy group were also synthesized. Although differences in the mass intensity ratios and retention time were expectedly smaller than with the side chain isomers, the identity of 4-CP could reliably be established.

It was only then that we realized that 4-CP is consumed in considerable quantities as a pharmaceutical to lower the lipid concentration in blood; biological tests showed that 4-CP has no herbicidal activity [12]. The biological tests were carried out as described in chapter 16.

Therefore, its was assumed as working hypothesis that the contamination found in the ground water and also the tap water originates from regular medical treatment via sewage. The prerequisite for checking this working hypothesis was the development of a highly sensitive analytical method applicable to all types of water from sewage to tap water. This was achieved with the target compound analysis using time programmed SIM as described in the former section.

5.9.5 Automated Screening Applying Full Scan Acquisition

In earlier sections of this chapter it was demonstrated that residue analysis in water samples at the concentration levels of 100 or 200 ng per liter cannot be reliably performed with GC-MS in full scan mode. Only a few pesticides show mass spectra suited to library search. Two approaches to tackle this problem were described: 1. Screening analysis with selected detectors exhibiting higher sensitivity, such as ECD, NPD, FPD and AED followed by dedicated confirmation with GC-MS in SIM mode; 2. Target compound analysis with time programmed SIM window setting with the naturally imposed restriction on the number of monitored pesticides.

Both methods, have severe disadvantages. The combination of screening with selected detectors and additional confirmation of suspicious peaks is suitable in monitor tap water and ground water samples containing little matrix. This combination is not feasible with water of high matrix burden such as river water or sewage water. This holds true for the popular application of the ECD, while the situation is much improved using the AED.

The application of target compound analysis with SIM exhibits the inherent limitation to the pesticides selected as targets for the monitoring. Since the method is transparent to all other contaminants it may occur that while detecting the residue of one pesticide at the low ng per liter concentration level another heavy contamination in the µg or even mg per liter level is missed because the particular pesticide is not included in the analytical method.

More than 400 pesticides can be analyzed with GC-MS and many of them are extracted with good or medium recovery with LLE or SPE. These pesticides can all be detected in a gas chromatogram if a suitable detection method is applied. Such a universal detection method is undoubtedly MS when operated in full scan.

Therefore, in recent years, the mass spectrometer which has been used in confirmatory analysis for decades in the field of environmental analysis has also gained popularity in screening analysis. This new application may become a domain of the mass selective detectors or other simple quadrupole devices. In the course of the investigation of a series of ground water samples, we found that most of the chromatograms obtained by GC with ECD could not be evaluated satisfactorily, because of the great number of matrix peaks. Screening analysis with the HP 5970 mass selective detector (MSD) with cyclic scanning and searching in a designated mass spectral pesticide library showed surprisingly good results. Manual data evaluation still remained very time-consuming, although the search was only carried out in a designated library. Therefore, the macro program AUTOEVAL was developed which simulated automatically all the steps usually applied in manual data evaluation [48]. The program has been used successfully during the last few years with pesticide residue analysis in food and various kinds of environmental analysis in ground and surface water as well as soil samples. Recently an upgrade of this macro program was published with the new name AUTARG referring to the feature of an additional screening for specified target pesticides in the raw data of the full scan acquisition [49]. AUTARG, taking full advantage of WINDOWS, follows the line an analyst would take in evaluation the mass spectral data acquired. Such an evaluation includes autointegration with automated peak recognition, background subtraction and library search in designated pesticide libraries and also in universal mass spectral libraries. The decision concerning the presence of a pesticide is supported by quality factors and needs final inspection by the analyst. A very important feature is the use of retention times with pesticide recognition as a very important independent piece of information. This enables the recognition of target compounds overlapped by matrix compounds, which produce poor library search results. On the other hand, any similarity between the mass spectrum of the sample peak and the reference spectrum must be considered as purely coincidental if the retention times of the reference target compound and sample peak are significantly different. AUTARG provides, when operated on its first level, a complete analysis report with all integrated peaks listed with their retention times and search results. In a second result table,called Important Peak List, only those results of the library search are compiled which have met user defined thresholds for hit quality and retention time windows. For example, this list might contain only search results with a hit quality better than 80 and also all peaks with a hit quality less than 80, if exhibiting a difference in retention time smaller than one minute from that laid down in the library for the suggested compound. A third result table contains the integration results of all peaks with all such details as peak area, peak width, resolution, peak start and peak end. Together with the TIC picture, the integration result table presents an overview of all compounds detected in the sample as shown in Fig. 22.

All results are printed to files, which speeds up the data analysis process considerably in comparison to printing on paper. Data processing consumes less time than needed by the GC/MSD system to cool down after a run with temperature program and to equilibrate before the next start. A second advantage of AUTARG Level 1 is that the analyst can check the results first on the screen and can decide later what he wants to print out. This saves printing costs and reduces paper consumption enormously. Finally, the automatic comparison of search results with and without background subtraction always guarantees the best search results.

As mentioned above, the AUTARG Level 1 results table may contain some search results with a poor hit quality but excellent correlation in the retention

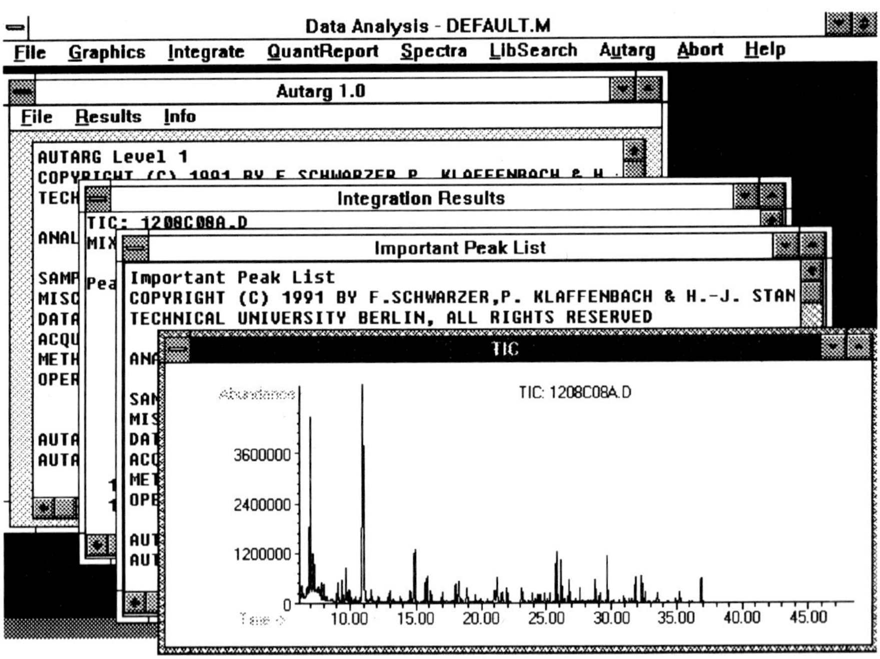

Fig 22. Results of a AUTARG Level 1 search (Reproduced with permission from Ref. [49])

times for the peak searched and the suggested library compound. In this case, the target compound may be overlapped by a co-eluate from the matrix and manual evaluation had, therefore, to be performed. Such a manual verification procedure, performed to confirm the identity of a compound generating a peak overlapped by the peak of a matrix compound, has been demonstrated with the case story of the recognition of clofibric acid in ground water. This process of manual evaluation of peaks exhibiting poor hit quality to suggested library compounds, but a good correlation in retention time can provide excellent results when using RIC of appropriate selected ions. A disadvantage of this technique is that it is very time consuming when applied to a great number of peaks. Therefore, AUTARG Level 2 was developed to execute automatically all the steps described above. The user has to create so called 'control files' that define two indicative ions and the time window for the target compounds.

Level 2 begins to check for the first target compound by redrawing the specified ion traces in a user-defined time window, usually 1 minute, around the expected retention time of the target compound. Only if Level 2 has found peaks in the traces of both ions will it proceed to check whether peaks appear at the same retention time in both of the ion traces. If their difference in retention is smaller than 0.015 minutes, Level 2 assumes that the two ions originate from the same target compound and continues by taking a scan at the apex of the peak followed by background subtraction. This is performed by subtracting the scans taken at peak start and peak end. A target compound library is searched, to find a match for the resulting spectrum. It can usually be seen when looking at the RIC traces that the apices of the target compound ion traces do not fit exactly to the apices of the matrix compound peak. Therefore, the spectrum of the scan selected by Level 2 generally shows already a good correspondence to that of the target compound sought for. Further background subtraction at peak start and peak end of the target compound may eliminate the interfering ion almost completely, so that the library search now results in a better hit quality. Library search is also performed without background subtraction. Only the better result is reported. Level 2 then continues to search for the next target compound listed in the control-files that have a total search capacity of up to 500 target compounds.

A particular feature of Level 2 is also based on RIC-technique. Level 2 can find such peaks that are normally overlooked being hidden in the noise. These peaks naturally are not integrated by the integration software, and as a consequence also not sought. Since the signal to noise ratio is, with most of the ion traces, orders of magnitudes better than with TIC, the presence of compounds can be spotted at very low concentration levels depending, of course, on the overall abundance of the fragment ions selected. This is demonstrated with the example below.

Figure 23 shows the TIC of a ground water sample where Level 1 reported no search results. With Level 2 using a 'control file' for phenoxyalkanoic acid PFB esters, 5 target compounds were recognized as indicated in the legend of Fig. 23.

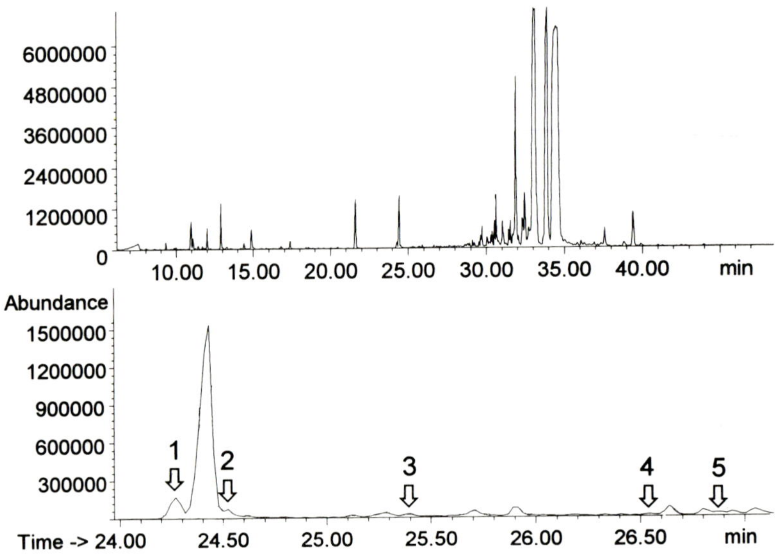

Fig 23. TIC of a ground water sample with AUTARG Level 2 search for phenoxyalkanoic acid herbicide PFB-esters *top*: complete chromatogram; *bottom*: zoomed part with the peaks labelled which were found by AUTARG Level 2 (Reproduced with permission from Ref. [49])

Peak	Target Compound	Hit Quality
1	clofibric acid-PFB-ester	98
2	2,4-dichlorobenzoic-acid-PFB-ester	96
3	mecoporp-PFB-ester	83
4	MCPA-PFB-ester	42
5	dichlorprop-PFB-ester	52

AUTARG was tested with three groups of target pesticides at the relevant concentration levels of 100 ng/l [50]. Standard mixtures containing 18 chlorinated pesticides, 16 triazine herbicides and 21 phenoxyalkanoic acids were subjected to GC-MS analysis in full scan applying a HP MSD (5970) mass selective detector. 2 µl of the solutions in toluene were injected hot splitless containing 2 ng of each pesticide. All pesticides were recognized with an AUTARG Level 1 search with the exception of two, namely metribuzin and triadimefon demonstrating a satisfactory performance of the instrument. Metribuzin was recognized with AUTARG Level 2 search with a quality of 40 at the calibrated retention time. After spiking a ground water sample with the pesticides to achieve a concentration of 100 ng/l, 1 l of the sample was extracted once with 50 ml dichloromethane to recover chlorinated pesticides and triazine herbicides. Phenoxyalkanoic acids were extracted using SPE with RP-C18 adsorbent after acidification. All extracts were concentrated to 100 µl and 2 µl injected for GC-MS analysis. Five out of 18 chlorinated pesticides were recognized from the water extract with AUTARG Level 1, but 12 with AUTARG Level 2 search. The two most volatile chlorinated pesticides were lost due to the lack of keeper addition. Heptachlor and quintozene were not identified because of their poor mass spectral characteristics (heavy fragmentation), p,p'-DDT due to its susceptibility to the chromatographic conditions and mirex as the last eluting compound also with widespread fragmentation. With the triazine herbicides 10 and 11 of the 16 target compounds were recognized with better quality of the matches obtained with AUTARG Level 2 search. None of the 21 phenoxyalkanoic acids could be found in this investigation with AUTARG Level 1 search but 7 with Level 2.

Our conclusion was that automated pesticide residue analysis can be reliably performed by means of the support of AUTARG. Level 1 is a valuable and reliable tool for the automated evaluation of GC/MS data. It replaces the time consuming manual evaluation by providing similar reliable results. It can be performed right after data acquisition using the time elapsed before the next injection.

AUTARG Level 2 is a powerful addition to Level 1, especially in trace level analysis, when looking for specific compounds by using dedicated control files. It has been proved that the use of ion traces of Level 2 makes the detection of target compounds hidden in the chromatographic background possible. This is achieved because the signal to noise ratio is much better than that of a TIC-chromatogram, and secondly, because the ion traces suppress overlaying matrix compounds, so that hidden target compound peaks can be found.

These investigations using an older GC/MSD system have furthermore shown that the limits of AUTARG are given by the instrumental limits of detection. New GC-MS systems promise to perform automated evaluation of full scan GC-MS analysis of water samples according to the tolerances for drinking water of the European Community for an increasing number of active compounds.

6 References

1. Brauch H-J, Schullerer S (1991) Fresenius J Anal Chem 339: 420
2. Brauch H-J (1993) Acta hydrochim hydrobiol 21: 84
3. EPA-Methods, Method 525 Determination of organic compounds in drinking water by liquid-solid extraction, capillary column gas chromatography/mass spectrometry
4. Ferderal Register/Vol. 49, No. 209/Friday, October 26, 1984/Rules, Regulations Methods 625- Base/Neutrals, Acids
5. EEC drinking water guideline, 80/779/ EEC No. L229/11-29, EEC, Brussels, August 30th, 1980
6. Jimenez NC, Atallah YH, Bade Th R (1989) J Assoc Off Anal Chem 72: 840
7. Lee H-B, Peart ThE, Carron JM, Tse H (1991) J Assoc Off Anal Chem 74: 835
8. Adolfsson-Erici M, Renberg L (1991) Chemosphere 23: 845
9. Legler J (1991) VDLUFA, Section XI, "Environmental-Analysis" Fresenius J Anal Chem 339: 417

10. Wittmann R, Weber WH, Baker LIT 225
11. Koterba MT, Banks WSL, Shedlock RJ (1993) J Environ Qual 22: 500
12. Stan H-J, Linkerhägner M (1992) Vom Wasser 79: 75
13. Muir DCG, Segstro MD, Welbourn PM, Toom D, Eisenreich SJ, Macdonald CR, Whelpdale DM (1993) Environ Sci Technol 27: 1201
14. Haberer K, Oemichen U (1986) Vom Wasser 66: 225
15. Lartiges S, Garrigues P (1993) Analusis 21: 157
16. Nations BK, Hallberg GR (1992) J Environ Qual 21: 486
17. Stan H-J, Bockhorn A (1991) Fresenius J Anal Chem 339: 158
18. Barcelo D, Sole M, Durand G, Albaiges J (1991) Fresenius J Anal Chem 339: 676
19. Klaffenbach P, Dethlefs F, Butz S, Stan H-J (1991) Determination of Carbamate and Urea Pesticide Residues in Food and Water Samples Applying Multiresidue Methods for Clean Up. Screening with AMD-TLC and Confirmation with Capillary GC-MSD-Poster presentation on the 13th International Symposium on Capillary Chromatography, May 13-16th (1993), Riva del Garda
20. Bolygo E, Atreya NC (1991) Fresenius J Anal Chem 339: 423
21. Köhle H, Haberer K (1990) Vom Wasser 75: 75
22. Hadfield ST, Sadler JK, Bolygo E, Hill S, Hill IR (1993) Pestic Sci 38: 283
23. Klaffenbach P, Holland PT (1993) J Agric Food Chem 41: 396
24. Robbat A Jr, Liu C, Liu T-Y (1992) J Chromatogr 625: 277
25. Karg FPM (1993) J Chromatogr 634: 87
26. Farber H, Scholer HF (1993) J Agric Food Chem 41: 217
27. Tsuchiya Y, Ohashi N (1992) Jpn J Toxicol Environ Health 38: 560
28. Foster GD, Foreman WT, Gates PM (1991) J Agric Food Chem 39: 1618
29. Mattern GC (1991) Diss Abstr Int B 52: 1155B
30. Kadokami K, Morimoto M, Haraguchi K, Koga M, Shinohara R (1991) Anal Sci 7: 247
31. Johnson WE, Fendinger NJ, Plimmer JR (1991) Anal Chem 63: 1510
32. Clark K, Gomme J, Hennings S (1991) Pestic Sci 32: 15
33. Bagheri H, Vreuls JJ, Ghijsen RT, Brinkmann UTA (1992) Chromatographia 34: 5
34. Mattern GC, Louis JB, Rosen JD (1991) J Assoc Off Anal Chem 74: 982
35. Mills MS, Thurman EM, Pedersen MJ (1993) J Chromatogr 629 11
36. Schuette SA et al (1990) Analytica Chimica Acta 236: 141
37. Noij ThHM, van der Kooi MME, van Leerdam JA (1993) GC/MS screening of polar pesticides in water- a quantitative approach-Poster presentation on the Fifteenth International Symposium on Capillary Chromatography, May 24-27th (1993), Riva del Garda
38. Benfenati E et al (1990) Chemosphere 21: 1411
39. Brauch H-J, Schullerer S (1992) Bestimmunung von polaren PBSM-Wirkstoffen in Grund-und Trinkwasser mittels GC/MSD nach Derivatisierung In: Weber R, Weber E (eds) Buch der Umweltanalytik Bd4-Methodik und Applikation in der Kapillargaschromatographie. GIT Verlag Darmstadt, p 83
40. Schlett C (1990) Z Wasser-Abwasser-Forsch 23: 32
41. Peldszus S et al (1990) Vom Wasser, 75: 35
42. Entwurf DIN 38407, Teil 14 Dezember (1990)
Bestimmung von Phenoxyalkancarbonsäuren mittels Gaschromatographie und massenspektrometrischer Detektion nach Fest-Flüssig-Extraktion und Derivatisierung (F14) in: Deutsche Einheitsverfahren zur Wasser-, Abwasser-und Schlammuntersuchung
43. Weber W, Baker Lit 223
44. Färber H, Peldszus S, Schöler F (1991) Vom Wasser 76: 13
45. Infante R, Pérez C (1991) Intern J Environ Anal Chem 43: 165
46. Wittmann, R, Weber, W (1990) Lebensmittelchemie 44: 104
47. Haberer K et al (1991) Vom Wasser 76: 287

48. Klaffenbach P, Stan H-J (1991) J High Resolut Chromatogr 14: 754
49. Stand H-J, Schwarzer F (1993) J Chromatogr 653: 45
50. Stan H-J, Heberer Th (1993) J Chromatorgr 653: 55
51. Heberer Th, Butz S, Stan H-J (1994) JAOAC lntern 77: 1587
52. Butz S, Heberer Th, Stan H-J (1994) J Chromatogr A 677: 63
53. Heberer Th, Butz S, Stan H-J Intern J Environ Anal Chem 58: 43
54. Brunnee C (1987) Int J Mass Spectrom Ion Proc 76: 125
55. Biller JE, Biemann K (1974) Anal Lett 7: 515
56. Stan H-J, Lipinski J (1989) HP Pesticide Library, Hewlett Packard, Palo, Alto, CA
57. Holland PT (1990) Pure & Appl Chem 62: 317
58. Budzikiewicz H (1992) Spektren wichtiger Lösungsmittel sowie von Säulenuntergrund. In: Massenspektrometrie, VCH, Weinheim, p 165

Biochemical Methods

K. Rubach

Institute of Food Chemistry, Technical University Berlin, Gustav-Meyer-Allee 25, D-13355 Berlin, FRG

The outlines of this chapter are focussed on biochemical assay reactions and are marked off strictly from the so-called bioassays. The theory and principles of the immunoassays used are described such as the antigen-antobody reaction, the antibody production, factors influencing the assays. Examples of application to the analysis of pesticides in waters are cited as well as commercial immuno-kits for this purpose are listed. Besides this, methodes for the trace analysis of pesticides by enzyme inhibition as well as the use of reactions between the analyte and biosensors are discussed.

Chemistry of Plant Protection, Vol. 11

1 Introduction

The biochemical screening analysis of ground and surface waters includes immunoassays, enzyme inhibition tests and the biosensor technique. The basics of the antigen-antibody reaction, and the development and optimisation of immunoassays are briefly described. Although the reagents used in immunoassays are antibodies, and therefore of biochemical origin, they are not used as components of a so called ‘bioassay’. This article differentiates therefore strictly between the biochemical assay and the bioassay. Only the former are subject matter for the survey presented (One of the latter type is presented in Chapter 16). The same holds true for enzyme inhibition tests. Biosensors, however, can be prepared using enzymes or antibodies, as well as by employing living cells from micro-organisms or algae.

The immunoassays present with their excellent sensitivities and specificities a valuable supplement to the usual chromatographic-spectrometric methods for screening analysis. They allow a high sample throughput at reasonable cost.

Whilst the immunochemical methods have taken a notable position beside established chromatographic-spectrometric methods, this is not yet the case for the enzyme inhibition tests and applications of biosensor systems.

It must be understood from the beginning that, although the reagents to be used – the antibodies (Ab) – are namely of biological origin, they are not themselves being used as components of a so-called bioassay.

It is clear that, for the development and evaluation of an immunoassay, the same generally valid criteria as for every other chemical-analytical measurement or detection procedure are to be applied. An understanding of the immunochemical methodology may be somewhat hampered by the unique jargon, whose individual terms may confusingly be not always clear.

A number of synopses from the last 10 years go into the nomenclature problem [1–3]. For the immunochemical assay, analytes and their reactants are, of course, generally defined just as in every other chemical reaction. The advance and scope of the reaction is set by a measurement which shows a clear relationship to the analyte concentration. The principle of enzyme inhibition is in its nature more severely limited in application relative to the immunoassay and indeed also more difficult to master. The number of practicable tests developed in this field in the period of the last 10 years remains small. Biosensor systems for the screening analysis of pesticides in ground and surface water matrices have not, up to the present, achieved the breakthrough. Especially, the extremely low legally required detection limits cannot – excepting a few applications – be achieved.

2 Immunoassays

Immunoassays were formerly summarised in order to quantitate pesticides in plants within this book series (141). Here, communications are focussed on the objectives of water analysis. The term immunoassay is understood as a chemical analytical technique through which, making use of specific antibodies (Ab) produced by a biological process (immunisation), concentrations of analytes present (Antigen or Hapten) may be measured with high sensitivity and selectivity. Haptens are relatively small molecules with molecular weights of around a few hundred daltons. They themselves possesses no immunogenic properties and do not produce any immune response in the body of the experimental animal. In conjugation with a higher molecular weight protein, which is itself immunogenic, they effect a production of specific antibodies in the body of the experimental animal. In contrast, antigens, which are higher molecular structures with molecular weights above 5000 Da possess, as a rule, several structures, the so-called determinants, on their surface. Pesticides are, relative to their molecular size, so small that conjugation must take place in every case with a suitable protein before actual immunisation in order to establish the required immunogenity in the whole molecule.

2.1 Antigen-Antibody Reaction

The antigen-antibody reaction forms the basis of all immunoassays. It is an example of a reversible reaction and may be described by the Law of Mass Action:

$$\mathrm{H} + \mathrm{Ab} \rightleftharpoons \mathrm{HAb}$$

$$\mathrm{K} = \frac{[\mathrm{HAb}]}{[\mathrm{H}]\,[\mathrm{Ab}]}\ \mathrm{L\,mol^{-1}}$$

In the presence of a constant antibody concentration, possessing a particular affinity, the total hapten (analyte) quantity determines the relationship of bound to free hapten (b/f). In reverse, the relationship b/f in a sample allows the determination of the actual hapten concentration.

The dose-response curve is described by the following equation, valid for all immunoassays and independent of any special type of signal generation [4].

$$(\mathrm{b/f})^2 + (\mathrm{b/f})(\mathrm{KH_T} - \mathrm{KAb_T} + 1) - \mathrm{KAb_T} = 0$$

b/f being the relationship of bound to free hapten
H_T the total hapten concentration
Ab_T the total quantity of antibody and
K the reaction or affinity constant.

The course of the function $b/f = y[H_T]$ is hyperbolic. Since the relationship b/f cannot be directly measured in analytical practice, the additional introduction of a tracer is necessary in the case of a competitive immunoassay. The tracer differs from the hapten in that it is covalently bonded with a signal generator. The original immunochemical properties of the hapten, particularly its structure in relationship to the binding site on the antibody surface should remain unchanged.

With a constant deficit in antibody concentration and a given tracer concentration, the higher the hapten concentration in the sample happens to be then the lower the quantity of bound tracer will be.

Figure 1 shows for example the sigmoid courses of two dose-response (standard) curves for the determination of the herbicide mecoprop, respectively its methyl ester through a heterogeneous competitive enzyme immunoassay (ELISA), making use of polyclonal antibodies from rabbits and a horseradish-peroxidase conjugate as tracer.

The analyte concentration is taken from the logarithmically scaled abscissa. The curvilinear form of the standard curves improves the evaluation of the measurements. A representation often chosen is the form logit vs log c. Logit is here

$$\text{logit}\,\frac{B}{B_0} = \log \frac{\frac{B}{B_0}}{1 - B/B_0}$$

and log c the logarithm of analyte concentration. This term represents a straight line with negative slope.

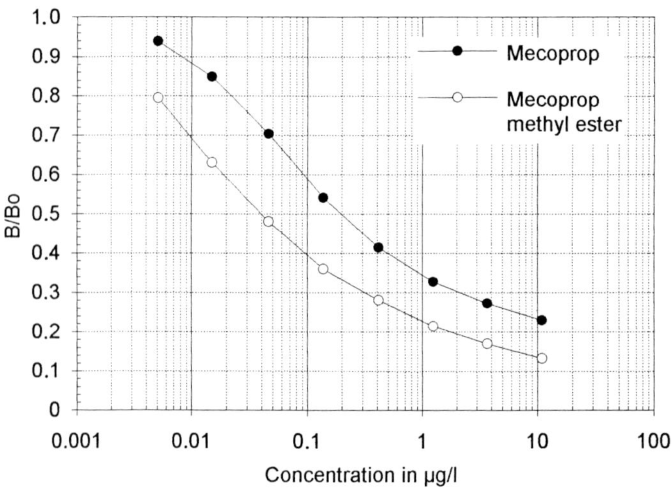

Fig. 1. Standard curves for the determination of the herbicide mecoprop and its methyl ester

B being here the measured value of the individual sample, for example an extinction value in the case of an enzyme immunoassay, and

B_0 being the value for the hapten concentration zero.

2.2 Antibody Production

The antibodies brought into use for the immunoassay determine the selectivity (specificity) of the test, and are additionally decisive along with the selected signal generator through the size of the affinity constant, for the sensitivity, respectively the detection or determination limit.

Specific antibodies can be produced in three ways:

1. Production in the bodies of animals (vertebrates)
2. Production with the help of Hybridoma techniques
3. Production with the help of Transgen techniques

The production of effective immunogens is an important aspect in the development of immunochemical assays. The lymphocytes in the vertebrate body involved in immunisation only recognise substances applied intradermally, subcutaneously, intramuscularly or intravenously as foreign when they exceed a size (molecular weight) of around 5000 Da. Lower molecular weight pesticides, excepting a few pyrethroid insecticides are therefore not immunogenic. By binding them to larger molecular structures, most of all proteins - bovine serum albumin, human serum albumin, egg albumin, keyhole limpet haemocyanin, thyroglobulin and fibrinogen coming as such into consideration, the immunogenic effect of the whole molecule will be established. The specific antibodies arising as immune response now also recognise the hapten in its unconjugated form. The synthesis of the immunogen is the only possible way to influence the selectivity (specificity) of the serum to be created. It is clear that antibodies will not only be formed against the hapten in the immunogen molecule. On producing an antiserum, care must be taken in the choice of carrier protein such that any protein structure present in the analyte matrix cannot cross react with the antibodies. There is a covalent bond between hapten and carrier protein. It should be sufficiently stable relative to hydrolytic influences under the conditions of the immunisation. The perfect hapten (immunogen) should contain as much information as possible about the actual target compound (analyte). It is, however, not always a requirement that the hapten structure be identical with the chemical structure of the analyte. To avoid steric hindrance on the spatially extensive surface of the carrier protein, haptens are best bound to the protein by means of spacer arms. Suitable spacers are alkane chains with a length of 3–6 carbon atoms, which should therefore be of low polarity and without additional substitution, but display terminally a suitable functional group, capable of reaction, such as $-NH_2$, $-COOH$, $-OH$, or $-SH$. Although no exact, detailed conception exists of how the antibody population present in a polyclonal antiserum actually "recognises", one may presume that as well as the provision of a suitable spatial distribution (fit), essentially strength effects at the non-covalent bond between hapten and antibody are involved. The same is valid for the bonding of the immunogen to the receptor of the β-lymphocytes. Decisive for the strength of the bond and hence for the affinity between antigen and antibody is the formation of hydrogen bonds, electrostatic interactions, van der Waal's forces and hydrophobic interactions. The affinity between antigen and antibody can be influenced by changes in pH value, ionic strength and reaction temperature. The highest selectivity (specificity) for the created antibodies occurs in general relative to the structural elements of the haptens which are furthest from the binding site of the carrier protein. Thus one has the means, by choice of binding site in the hapten part, respectively by skilled insertion of the spacer arm, to create either substance group-specific or substance-specific antisera. Although the choice of conjugation method depends mainly on the chemical characteristics of the hapten, and must therefore be individually defined, in practice, a number of basic techniques exist which have proven themselves for the synthesis of immunogens. Those methods have been reported in detail elsewhere [5–14].

Because the required reaction product is, in almost all cases, strongly contaminated with (low molecular weight) by-products, a subsequent cleanup by dialysis, ultra filtration, or suitable chromatographic and electrophoretic methods, e.g. size exclusion chromatography (SEC) and isoelectric focussing (IEF) is recommended. All enzyme immunoassay formats require either enzyme-marked haptens, or enzyme-marked antibodies. Synthesis of such compounds is achieved following the same principles as for immunogen synthesis.

There is no united opinion in the standard literature concerning the influence of the hapten density, that is to say the number of epitopes on the carrier protein, on the immunogenity of the conjugate. One may assume that the hapten density should be chosen to be neither too high nor too low. The coupling rate is measured for comparative examinations. This can take place either radiochemically, by marking of the hapten, by spectrophotometric measurement, or by titration of the amino groups which have not been converted. Other possibilities result from the nuclear magnetic resonance of phosphorus, fluorine, and nitrogen, so long as they are present in the hapten.

2.2.1 Polyclonal Antibodies

Antibodies are glycoproteins and belong to the class of γ-globulins. For radioimmunological and enzyme coupled measurements, antibodies of the G-class are used with a molecular weight of ca. 160 000 Da. The IgG molecule consists of two 'heavy' chains, each of 415 amino acids and two 'light' chains, covalently linked through two s–s bonds.

The binding sites for the hapten (antigen) are situated at the ends of the molecule with terminal nitrogen. The antigen determinants lie, with regard to the primary structure, in an area of variable amino acid sequence. Each IgG molecule possesses two binding sites, to which the immune-determinant sections of the antigens bind following the 'hand and glove' principle.

In order to produce the polyclonal antibodies, the immunogens, with the addition of so-called *adjuventeins*, are administered to the experimental animals. *Adjuventeins*, which contain dead mycobacteria in addition to mineral oil and detergents, stimulate the immune response. A too rapid resorption of the immunogen is simultaneously prevented. A reliable application plan does not exist. The type of injection, the immunogen quantity, and the chronological progression of the immunisation are determined on a completely empirical basis. The usual procedure is as follows, that the primary immunisation is followed, after a certain time, by a series of secondary injections, the so-called 'boosters'. The progress of the immunisation is checked by measuring the 'titres', which indicate the *avidity* of the antibodies to their antigens. The term 'titre' means the dilution of an antiserum for which, taking into consideration temperature, incubation time, buffer system, test volume and separation technique, in a clearly laid down test system, 50% of the tracer would be bound in the absence of non-marked antigens. As well as the titre, the affinity of the antibodies to their binding partners, the haptens, is an important quality measurement. One possibility of calculating the affinity constant is offered by the so-called Scatchard plot (15) of a serum.

The Scatchard plot is a graphic representation of the relationship

$$b/f = K([Ab_T] - B)$$

Whereby
b/f is the ratio of bound to free antigen,
K is the affinity constant of the binding protein,
Ab_T is the total amount of antibody, and
B is the concentration of the bound antigen.
It results in a straight line with negative gradient K.

The intersection with the b/f axis has a value of $K[Ab_T]$, that with the B axis of $[Ab_T]$. The determination of the affinity constant is only meaningful when an homogeneous antibody population is present. This is not the case for polyclonal antisera, but certainly so for monoclonal sera, which possess a uniform affinity.

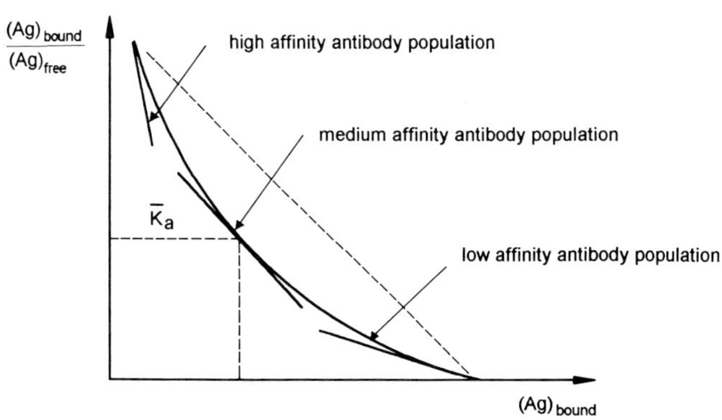

Fig. 2. Scatchard plot

The average intrinsic affinity (association) constant K_0 is measured for polyclonal sera. In practice, the function deviates to a greater or lesser degree from a straight line (broken line), an expression of the presence of antibodies in the antiserum with varied affinities. Antibodies with high affinities achieve affinity constants of $10^{10} - 10^{13}$ l mol^{-1}.

The so-called 'affinity chromatography' offers one possibility of separating high affinity immunoglobulins from the lower affinity antibodies, employing immobilised haptens. A detailed description of the methods of hapten immobilisation, and the elution conditions in the actual separation process are to be found in the publication by Tijssen [5].

In contrast to the antibodies appearing in polyclonal antisera, antibodies for pesticides and agricultural chemicals are to be considered as haptens compared to higher molecular weight compounds and because of their smaller size as relatively uniform concerning their complimentarity. When antisera are sufficiently strongly diluted, so that the interactions between the mostly avid immunoglobulins and their antigens are dominant, Scatchard plot often shows a great homogeneity regarding the antibody population. Antisera -possibly cleaned up with a suitable affinity chromatography method - are in general also adequate for the development of very sensitive assays.

2.2.2 Monoclonal Antibodies

Since the publication from Köhler and Milstein [16] concerning the production of monoclonal antibodies by single (isolated) clones of β-lymphocytes from the spleens of immunised animals, the so-called 'hybridoma technology' has attained great importance, as indicated by the number of publications within this branch of immunology.

Following fusion of the b-cells with the potentially immortal tumour cells, and careful screening of the hybridoma cells, the latter can be isolated and bred further in cultures which bring together the ability to produce antibodies of higher and uniform specificity with the possibility of unlimited duplication. The cross-reactivity for monoclonal antibodies (mAb) with respect to multivalent antigens, such as proteins, is notably decreased. Monoclonal antibodies show, relative to polyclonal sera, a tendency to lower affinity. Polyclonal and monoclonal antibodies show, in general, similar specificities toward haptens of limited size. By the further cultivation of the appropriate cell line, monoclonal antibodies can be produced in almost unlimited quantities with constant quality (characteristics). Although the hybridoma technique has become a routine method for appropriately equipped laboratories, the investment and running costs are relatively high in relation to the production of polyclonal sera.

An interesting application for monoclonal antibodies has been described by Langone et al. [17]. The authors reported production of stereo specific monoclonal antibodies for the determination of cotinine with an *idiotype-anti-idiotype-hapten-immunoassay*. Idiotype (Ab_I) resulted thereby from the immunisation of experimental animals with a synthetic hapten conjugate. Ab_I is then used as immunogen for the generation of the anti-idiotype (Ab_{II}), which is specific for the binding position of the hapten. Ab_I and Ab_{II} can be generated by the use of hybridoma technology with rigorous screening of the relevant clone. The original hapten then inhibits the reaction, in assay, between idiotype and anti-idiotype. Thus the necessity of producing marked hapten derivatives or macromolecular conjugates for binding in solid phase assays does not apply.

The production of antibody binding fragments in genetically engineered bacteria has recently been reported [18, 19]. Advances in molecular biology and the use of the polymerase chain reaction (PCR) for the duplication of genetic material lead to their expression, for example in E. coli. In the authors' opinion, it will, in the foreseeable future, no longer be necessary to immunise animals, since the combinational repertoire of non-immunised animals may well contain genes for the production of antibody binding fragments of required specificity. Such antibody binding fragments differ in size and characteristics from the usual immunoglobulins. For example, the capability of immunoglobulins of type IgG to react with protein A is based on a carbohydrate moiety in the F_c region, which is not present in the antibody binding fragments.

2.3 Assay Development and Optimisation

Should antibodies with sufficient selectivity and affinity be available, then there exists the additional means of assay development and optimisation in the balancing of antibody and conjugate concentrations. The term 'conjugate' is taken to mean a combination of the actual target substance (analyte) and a signal generator (the tracer or label). Since it is a matter of dealing with pesticides as mainly low molecular weight compounds, the underlying assay reaction will mainly be competitive. All immunoassays, so long as there is an underlying competitive reaction, are based on the competition between marked tracers and unmarked antigens, or haptens, for a limited number of antibody binding sites.

The tracer type determines, in the end, the method of measurement and, together with the affinity of the antibody/hapten reaction, also the sensitivity of the determination. The number of suitable tracers is limited. There are, in the main, three groups of signal amplifiers which find application in the development of immunoassays, namely radioisotopes, enzymes, and luminescent or fluorescent substances. Historically, the radio immunoassays [20] belong to the first developments of this type. Although this is a precise and sensitive technique, the several disadvantages cannot be overlooked. Radioactive tracers often have a limited shelf life. Too rapid decomposition, health risks and the demanding requirements relating to the provision of appropriate measuring devices, as well as the difficulties in disposing of the radioactive substances, led to the independent development of enzyme immunoassays (EiA) by Engvall and Perlmann [21] and Van Weeman and Schuurs [22]. Finally, fluorescent- and chemiluminescent-labelled methods have the potential for high sensitivity. A principle problem with this method is the high background signal during measurement caused by a large number of matrix components.

2.4 Diversity of Enzyme Immunoassays, Terminology

A large diversity exists in immunoassay formats. On these grounds, therefore, only a limited overview of the principle alternatives can be presented.

It is customary to differentiate between the so-called homogeneous and heterogeneous assays. In homogeneous assays, the analyte competes with the enzyme labelled analyte (conjugate) for the antibody. The binding of the antibody changes quantitatively the activity of the enzyme-analyte conjugate. A separation step is not required in this procedure. The sensitivity is lower than for heterogeneous techniques. The influence of the matrix on the enzyme activity often cannot be discounted. A prerequisite for the development of this method is a good knowledge of enzyme chemistry, since the covalent bond to the hapten should be spatially close to the active centre of the enzyme.

This assay has the acronym EMIT (enzyme activation and enzyme inhibition immunoassay). A separation step is unnecessary. The limited number of antibody binding positions are correlated to the number of the analyte molecules.

The higher the analyte concentration, the more pronounced is the modulation of enzyme activity.

The different methodologies for the so-called heterogeneous immunoassays are far more numerous than for the homogeneous methods. Heterogeneous assays require a step to separate the bound from the unbound tracer.

The immobilisation of the antibodies (Ab) on a suitable carrier, e.g. polystyrene microtitre plates, directly followed by competition between sample analyte and marked analyte for the antibody binding sites leads to signal measurement via a minimum number of steps.

Should the antibodies be easier to mark than the antigens, the latter may first be immobilised. The antigen molecules from the sample then react, along with the immobilised antigens, at the binding sites of the enzyme marked antibodies. By introducing an additional step into the assay procedure, enzyme marking of the antibodies may be dispensed with. The final detection is then effected with an enzyme labelled anti-immunoglobulin, which is, as a rule, commercially available.

An increase in sensitivity, but also with a simultaneous increase in assay reaction steps, comes into play when the antigen molecules are first immobilised on the plate. The analyte is then separately incubated with enzyme labelled antibodies. The mixture, which still contains free enzyme labelled antibodies, is then brought to react with the immobilised antigens. The unbound antibodies bind with the antigens on the plate.

This indirect ELISA can also be undertaken using marked anti-immunoglobulins which, when chosen correctly, can lead to a strengthening of the signal, though also to an increase in the number of reaction steps in the assay. There exists the possibility of increasing the reaction rate, namely by first incubating

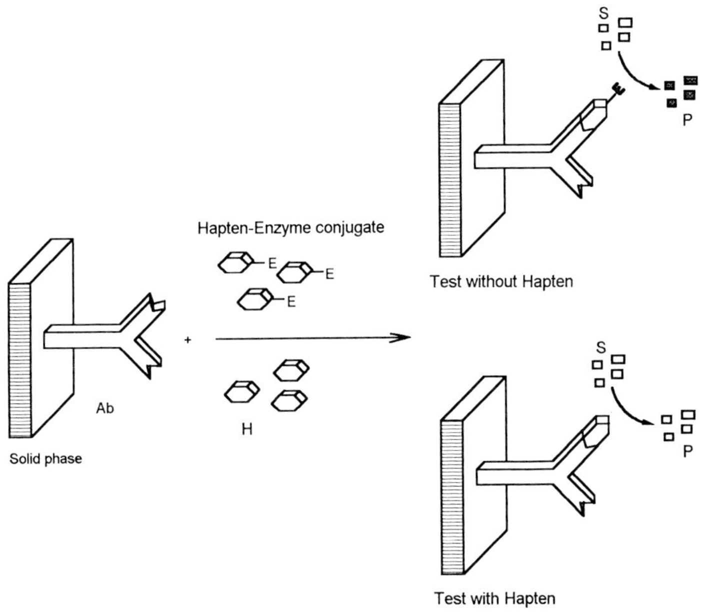

Fig. 3. Competitive Immunoassay

the analyte with immobilised antibodies. An excess of enzyme labelled analyte then reacts with the non-converted bound antibodies.

No immobilisation of antigens is required by the sandwich method, nevertheless a certain size of target molecule is a requirement. The analyte is incubated with immobilised Ab. Addition of the enzyme labelled Ab follows, and the Ag to be measured is, so to speak, surrounded. Should one choose specific antibodies from two different species, this can lead to an increase in assay sensitivity when an enzyme labelled anti-immunoglobulin is selected as the second antibody.

Competitive immunoassays offer better possibilites for the quantification of smaller molecules. Enzyme marked haptens and haptens compete for a limited number of immobilised binding sites on the antibodies. Subsequently, dependent on the substrate specificity of the enzyme, the conversion may be measured by choice of the suited substrate. The higher the concentration of the analyte, the smaller is the measured extinction or absorption.

Every competitive assay shows itself as an interaction between three parameters. These are the concentrations of hapten, antibody, and tracer. [23]

Whilst the experimentation with a given antiserum can have little or no effect on the affinity, the concentrations of antibody and tracer may be varied and optimised. The same holds true, of course, for other quantitative influences on the immune response, such as pH value of the reaction buffer, incubation time and temperature, and ionic strength to name but a few.

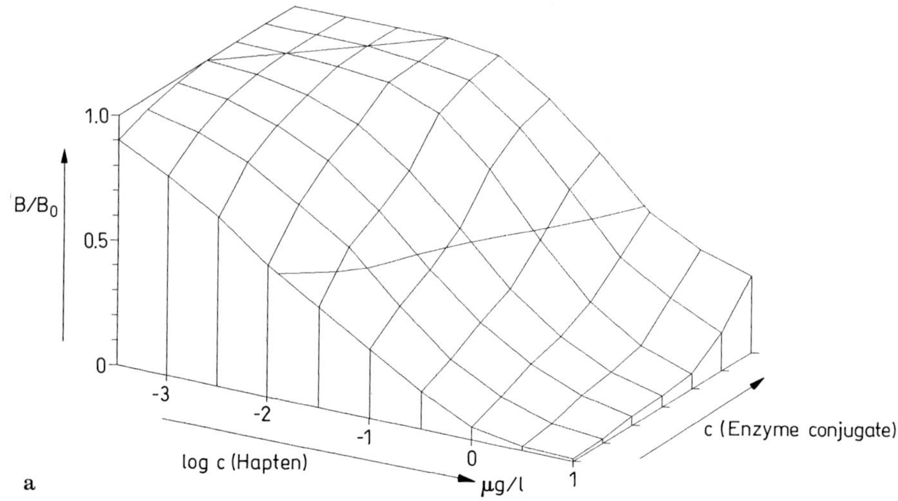

a

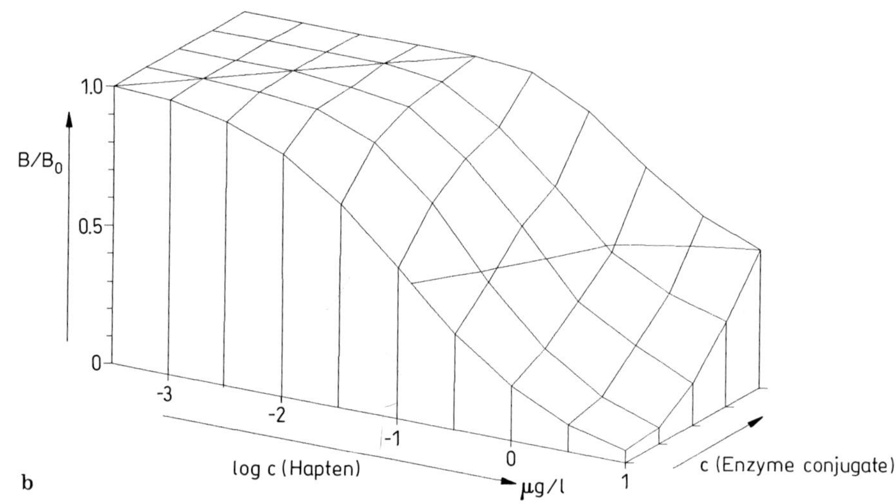

b

Fig. 4. Interaction between the concentrations of haptens, antibodies and tracers **a** for a lower antibody occupancy **b** for a higher concentration

Figure 4b shows the dependence of the immune response B/B_0 on the log of the hapten concentration and on the tracer concentration for a given (higher) antibody occupancy for the measurement of the herbicide mecoprop methyl ester. In the region of higher tracer concentration, the proportion of bound tracer changes little for a given, low hapten concentration. The detection limit increases. With further increases in antibody occupancy, even relatively high hapten concentrations are not able to displace the tracer, they merely occupy the free antibody sites. This leads in any case to a worsening of the detection limit. One can recognise from Fig. 4, that with an increase in antibody and tracer concentration, the detection limit in fact rises but the sensitivity, that is to say the gradient of the dose-response curve, is improved.

Figure 4a shows the interaction between hapten and tracer concentration for a particular (lower) antibody occupancy than in Fig. 4b. The surface horizontal to the hapten and tracer concentration axes is less well developed for lower antibody occupancy. Relatively low amounts of hapten may thus be detected with the appropriate low tracer concentration. The influence of an affinity constant with too low a value shows itself in an increased tracer concentration, to occupy all binding sites on the antibodies. This binding is, however, only reversible in character with higher hapten concentrations, meaning that the detection limit rises.

The optimisation of an assay is of great importance for the detection limit. Imperfect or non-optimised assays lead additionally to an excessive consumption of antiserum and tracer. Hock (23) summarised the equipment, time and expenditure involved in the development of an enzyme immunoassay based on polyclonal antibodies with s-triazine.

2.5 Evaluation of Accuracy and Precision

The precision of an immunoassay is, in contrast to other analytical techniques, following the sigmoid function of the standard curve, dependent on antigen concentration. The 'precision profile' offers a possible means of deciding on the optimum concentration region. In its simplest form, the coefficient of variation is calculated from a large number of measurements made with known analyte concentrations.

The working region can be established by defining the minimum acceptable precision (map). Experience shows in general that variation coefficients of below 10% are achievable when screening for pesticides using EIA methods. The function $CV = f(Ag)$, where CV represents the coefficient of variation, is the precision profile. Setting a minimum requirement for the CV function defines the working range. One must differentiate between 'within run' precision and 'between run' precision. The 'within run' precision is defined as the precision

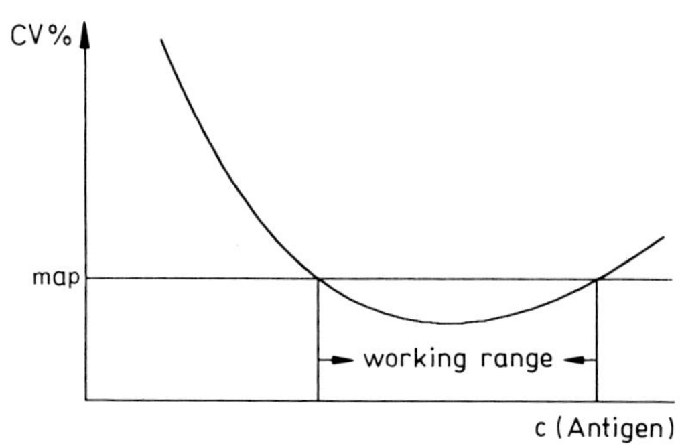

Fig. 5. Definition of the working range in the case of screening pesticides using EIA methods

of the identical sample, measured repeatedly by the same method. The 'between run' precision is a quantification of the assay's capability to reproduce a result in two separate runs with identical samples. Grounds for imprecision appearing are, of course, numerous and occur, in the end, because of the complexity of the complete test procedure. Even when the method design and the optimisation are outside the control of the user, such as with commercial assays, each single step offers ample possibilities for random and systematic errors. This relates both to the handling of the antibodies (binder), as well as to each form of separation during the assay procedure. The colour generating method and the preliminary measures, such as e.g. isolation, and extraction steps must also be included in these considerations. Additional influences come from the fact that commercial test kits are optimised for the shortest possible analysis time. Since achievement of exact equilibrium cannot be waited for, small deviations from the time plan may, under certain conditions, produce rather large fluctuations in the results. The same is also true for temperature accuracy, conditions related to the pH values of reaction and wash buffers, their ionic strengths, and so forth.

A further quality criterion for the immunoassay is its sensitivity. Understood under the term 'sensitivity' is the smallest amount of hapten to be measured which may be dependably differentiated from zero. This is a characteristic of the standard curve. The affinity of the antibodies, the specific activity of the tracer, and the suitability of the blank are simultaneously reflected in it. The sensitivity is thus both dependent on the gradient of the standard curve as well as on the precision of single measurements at or approaching the zero level. From this definition, it follows that for two assays with equal precision at zero, the steeper gradient of standard curve characterises the higher sensitivity. Conversely, for two assays with the same slope, the more precise one is the more sensitive. The least detectable dose (LDD) is generally measured from calculations of the mean value and standard deviation of the zero standard from repeated analyses. The mean value -6SD (occasionally also -3SD), where SD represents the standard deviation, expressed as measured value or in concentration units, is the smallest amount which diverges from zero with 95% certainty.

Both influential dimensions: the least detectable dose (LDD), and the (negative) slope of the standard curve, are brought together in the term sensitivity. With all assay variables set to their optimum values, the decisive value for the sensitivity is the affinity constant K_a. It is a measure of the binding energy between antigen and antibody in the antigen-antibody complex. With 50% saturation of available antibody binding sites, it follows that

$$[\mathrm{Ag}] \sim \frac{1}{K_a}$$

For high affinity antibodies ($K_a \sim 10^{10} - 10^{12}\ \mathrm{l\ mol^{-1}}$), the concentration of free antigen lies at $10^{-10}\ \mathrm{mol\ l^{-1}}$. According to Walker (24), the detection limit for an optimised radio immunoassay, also generally valid for an enzyme immunoassay, lies at $0.1K_a$.

Specificity is understood as the capability of an assay to produce a measurable response only for the analyte to be determined. An antibody is thus specific when it reacts exclusively with the target substance to be determined. Portions of the matrix should make no contribution to the response of the assay.

A quantitative measure of the specificity is the cross reactivity. This is understood as the reaction of an antibody with substances whose chemical structures are, to a greater or lesser extent, similar to that of the analyte. To determine the extent of the cross reactivity, so called displacement curves are recorded with the cross reacting substances to be tested, and thus the amount of cross reactant determined, which displaces 50% of the tracer. The extent of cross reactivity is calculated from the following relationship:

$$\text{Cross reactivity in }\% = \frac{\text{concentration of the hapten at 50\% B/Bo}}{\text{concentration of the cross reactive haptens at 50\% B/Bo}} \times 100$$

Displacement curves are often not parallel to the standard curve over the whole concentration range of the cross reactant. It is clear that cross reactions originating from the matrix can only be empirically established. As well as cross reactions from 'antigen-like' substances, the reaction rate is also influenced by experimental variables such as pH value, ionic strength, buffer composition, and organic solvents among others.

As well as the precision, the correctness limits the accuracy of an assay. The correctness of radio immunochemical and enzyme immunochemical determinations of higher molecular weight compounds, e.g. proteins and *proteohormones*, namely the ascertainment of the "true" value is only difficult to judge. For the determination of lower molecular weight compounds, to which pesticides may mainly be attributed, in addition to the availability of suitable standards, the choice of adequate confirmation procedures is of prime importance. Immunoassays deliver results which, as far as correctness is concerned, do not always fulfil all the analyst's wishes. The grounds for this lie in the condition that the means of production and the origin of the antibodies does not exclude cross reaction with substances of similar structures, although separation steps followed by *energy dispersing/resolving* detection contribute further to specificity. Immunochemical methods in environmental analysis for low m. wt. compounds are suitable for screening. Since false negative results rarely appear, but cannot basically be excluded, the actual importance of the immunochemical method lies in its combination with a spectrometric technique.

2.6 Applications

Following the early work of Ercegovich (25) and Centeno (26) on the immunochemical determination of DDT and malathion, a whole host of assays for the determination of pesticides, industrial chemicals and microbial toxins have been developed based on the antigen-antibody binding principle.

Advantages and disadvantages of immunochemical methods in comparison to normal spectrometric and chromatographic techniques have been repeatedly reviewed in detail (27–31), so that here a short acknowledgement of these aspects is considered as being sufficient.

The application of immunochemical assays to the determination and measurement of pesticides and environmental pollutants facilitates a high sample frequency, i.e. a large number of samples run in parallel. The ability to examine many samples in a relatively short analysis time – on average 1–3 h – without longwinded sample treatment, permits relatively low analysis cost, even taking the often considerable development costs into account. High sample throughput also allows data evaluation for the estimation of accuracy and precision.

By optimising them for speed, respectively short reaction time, immunochemical methods are excellently suited for analyses in the field. Optimisation with respect to sample number and precision allows immunoassays to be employed as so called "screening tests" in combination with the usual spectrometric techniques. Immunoassays generally function best in aqueous systems. Additions of polar solvents raise the solubility of the analyte, but, as a rule, bring no advantage for the antigen-antibody reaction.

Considering the molecular character of the analyte, the adoption of the immunochemical method is easier to put into practice the more hydrophilic (polar) the analyte is. This holds for all assay formats. The analyte molecules should also be relatively large and of low volatility, and should be sufficiently stable, under the conditions of the immunochemical analysis, to hydrolysis and other influences. Immunoglobulins (antibodies) which are employed in the immunochemical trace analysis generally possess sufficient affinity to the antigen (hapten). The affinity constants, depending on whether monoclonal or polyclonal antibodies are involved, lie in the region of $K_a = 10^{10} - 10^{12}$ l mol^{-1} and lend the test an adequately high sensitivity (test mean 0.1–1 $\mu g l^{-1}$). The pronounced specificity or selectivity of the antisera (antibodies) which come to be used in the immunochemical analysis presents equally advantages and disadvantages. High specificity, namely the capability of the immunoglobulins to react exclusively, or almost exclusively, with their antigens forming reversible, non-covalent bonds, allows the omission of a detailed clean up, such as is almost always required in the usual spectrometric methods. The appearance of cross reactions naturally limits the surety of immunochemical methods. Cross reactions cause a rise in the so called "false positive" results, and hence higher analysis costs relating to the thus necessary confirmation examination. On the other hand, the appearance of cross reactions caused by the presence of several analytes of similar structures makes the development of so called "group specific assays" possible, which then allow the detection or measurement of a limited number of compounds of similar structures, but unfortunately with varied sensitivities. For carefully optimised and validated tests, the specificity is high, i.e. the number of false negative results is low. This feature is the basis for the excellent reputation of immunoassays as screening methods. Immunochemical methods are generally not suited to the development of multi-analyte methods because of the inherent specificity of the immunoglobulins. In this respect, the immunochemical methods thus represent not a competitor, but possibly rather a welcome supplement to the established chromatographic and spectrometric techniques.

The low complexity of the apparatus, only a suitable special photometer being required for ELISA for example, and the high analysis frequency, make immunochemical methods most suitable for direct measurements in the field.

The most pronounced value of immunochemical methods of examination must lie in their application as screening methods. A reduction in the inherently high analysis costs proanalysis of chromatographic-spectrometric methods can in many cases be readily achieved by reduction of the extensive sample material.

The use of antibodies or antisera in water analysis, just as for other matrices in the field of foodstuffs, demands a detailed overall knowledge of interferences caused by the matrix. In any case, the presence of cross reactants should be tested for in detailed, thorough investigations. Doing without necessary confirmation investigations puts the value of immunochemical tests results in question. This holds for both the qualitative and quantitative aspects.

2.6.1 Home-Made Assays

The application of the antigen-antibody reaction to residue analysis has led to an extraordinarily large number of assay developments over the last two decades. It may be assumed that the actual number of developments – valid also for the assays listed in Table 1 – is considerably more than that expressed in the number of relevant publications. Especially the producers of pesticides and industrial chemicals may well have at their disposal further, unpublished assays for "in house" use. A distinction must be made between the so called "home made assays" whose development would mainly be in universities for use in basic research, and the commercially produced assays. Manufacture and sales of test kits suitable for residue examinations is closely connected with the problem of standardisation of such techniques and the realisation of the necessity of achievement of an official evaluation. Up to the present, well meant efforts in this direction have not been very successful in practice. The Environmental Protection Agency (EPA), the AOAC, and the Food Safety and Inspection Service (FSIS) have published guidelines for carrying out evaluation tests and trial series. (32–35)

Although Table 1 does not pretend to be complete, the assays presented reflect well the developments over the last 10 years.

Most of the assays in Table 1 have been specially developed for the water matrix (surface and ground water).

Diverse means of implementation come into consideration – microtitre plates, tubes, cards and beads, but the use of microtitre plates has gained the greatest importance. Difficulties with plate material homogeneity and reliability of the necessary measuring and evaluation equipment (reader) have been more or less overcome. The ELISA has asserted itself against the RIA in the last five years.

Table 1. Immunoassays for the determination of pesticides in ground and surface water

Pesticide	Class	Type of immunoassay	References
Benomyl	Fungicide	RIA, ELISA FIA	36, 37, 38
Fenpropimorph	Fungicide	ELISA	39, 40
Iprodione	Fungicide	ELISA	41
Metalaxyl	Fungicide	ELISA, EIA	42, 43
Thiabendazol	Fungicide	ELISA	38
Triadimefon	Fungicide	ELISA	88
Alachlor	Herbicide	RIA, ELISA	44, 45, 46, 47
Ametryn	Herbicide	EIA	48
Amitrole	Herbicide	ELISA	49
Atrazine	Herbicide	EIA, ELISA	50, 51, 48, 52, 53, 54, 55, 56 57, 58, 59, 60 61, 62, 63
Aziprotryn	Herbicide	ELISA	64
Bentazon	Herbicide	ELISA	65
Bromacil	Herbicide	ELISA	66
Chloroquinoline (derivate)	Herbicide	ELISA	67, 68
Chlorsulfuron	Herbicide	ELISA	69
2.4–D	Herbicide	RIA, EIA, ELISA	70, 71, 72, 73, 74, 75
2.4–DB	Herbicide	ELISA	74
Desethylatrazin	Herbicide	ELISA	76, 77, 78
Diclofop-methyl	Herbicide	FIA, EIA	79
EPTC	Herbicide	ELISA	81
Flamprop	Herbicide	ELISA	82
Fluazifop	Herbicide	ELISA	82
MCPB	Herbicide	ELISA	74
Mecoprop	Herbicide	ELISA	
Metazachlor	Herbicide	ELISA	83, 84
Metabenz-thiazuron	Herbicide	ELISA	85, 86
Metolachlor	Herbicide	EIA	42, 87
Molinate	Herbicide	ELISA	88, 89, 90, 91, 92
Norflurazon	Herbicide	EIA	93, 94
Paraquat	Herbicide	RIA, ELISA	95, 96, 97, 98 99, 100, 101, 102,
Picloram	Herbicide	EIA	83, 42, 84, 85 103, 42, 104, 105
Prometryn	Herbicide	ELISA	106
Simazine	Herbicide	ELISA	54, 95
Oryzalin	Herbicide	ELISA	107
Terbutryn	Herbicide	ELISA	108, 109
Thiobencarb	Herbicide	ELISA	81
Triadimefon	Herbicide	ELISA	110

Table 1. (continued)

Pesticide	Class	Type of immunoassay	References
Triazines	Herbicide	FIA, ELISA	111, 112, 113, 56 114, 115, 116, 117, 118, 119, 95
2.4.5.-T	Herbicide	RIA	70
Aldicarb	Insecticide	ELISA, EMIT	120
Aldrin	Insecticide	RIA	121
Alsystin	Insecticide	ELISA	122
S-Bioallethrin	Insecticide	RIA, ELISA	123, 124
Cyclodienes	Insecticide	ELISA	125
Cypermethrin	Insecticide	ELISA	126
DDT	Insecticide	Immundiffusion	127
Diflubenzuron	Insecticide	ELISA	107, 122, 128, 129
Dieldrin	Insecticide	RIA	121
Malathion	Insecticide	Immundiffusion	127
Paraoxon	Insecticide	RIA, ELISA	130, 131, 132
Parathion	Insecticide	RIA, ELISA	130, 131, 132
Permethrin	Insecticide	ELISA	
Pyrethroids	Insecticide	ELISA	133
Warfarin	Rodenticide	RIA	134

The difficulties in handling and disposing of radioactive materials gives ELISA the advantage. The greater number of new developments are concerned with heterogeneous assays following the competition principle. Since analytes from the field of pesticides are, as a rule, relatively small (< 1000Da), the sandwich technique may be ignored. That is also valid, for different reasons, for the development of so called homogeneous assays implemented in the form of EMITS, which bring certain advantages on the grounds of shorter analysis time (minutes) but must be seen in a less positive light when considering matrix interferences.

As relates to the detection limit, due to the generally strict legal requirements, a test mean in the region of 0.1 μg/l is demanded. Most new developments for pesticide determination in the water matrix fulfil this requirement, in a few cases they even lie markedly below.

2.6.2 Commercially Produced Assays

Table 2 shows a perhaps inevitably incomplete list of commercially produced immunokits for the detection and determination of pesticides. As may be gathered from the table, the detection limits for many kits do not lie adequately below the maximum levels for drinking water set by the legislatures of many individual countries. Concerning the question of assay format, it should be noted that development in home made assays is reflected in commercially produced tests. The assay in microtitre plate format with an enzyme tracer seems to be gaining ground on the other (older) developments. The use of radioactive tracers clearly does not

Table 2. Commercially produced immuno-kits for the detection and determination of pesticides in water

Analyte	Producer/Distributor	Antiserum	Assay format	Detection limit*
Alachlor	AVENA GmbH Markt Schwaben, Germany	polyclonal	microtiterplate tubes	0.05 $\mu g\ l^{-1}$ 0.1 $\mu g\ l^{-1}$
	J. T. Baker B.V., Groß–Gerau, Germany	polyclonal	magnetic particles	0.05 $\mu g\ l^{-1}$
	Immuno Systems, Inc., Scarborough, ME, USA	polyclonal	tubes	0.5 $\mu g\ l^{-1}$
Aldicarb	AVENA GmbH Markt Schwaben, Germany	polyclonal	microtiterplate tubes	0.4 $\mu g\ l^{-1}$ 0.3 $\mu g\ l^{-1}$
	J. T. Baker B.V. Groß Gerau, Germany	polyclonal	magnetic particles	0.25 $\mu g\ l^{-1}$
Atrazin	J. T. Baker B.V. Groß Gerau, Germany	polyclonal	magnetic particles	0,040 $\mu g\ l^{-1}$
Benomyl	AVENA GmbH Markt Schwaben, Germany	polyclonal	microtiterplate tubes	0.1 $\mu g\ l^{-1}$ 0.2 $\mu g\ l^{-1}$
	J. T. Baker B.V., Groß Gerau, Germany	polyclonal	magnetic particles	0,1 $\mu g\ l^{-1}$
Benomyl	Immuno Systems, Inc., Scarborough, ME, USA	polyclonal	tubes	0.4 $\mu g\ l^{-1}$
Captan	AVENA GmbH Markt Schwaben, Germany	polyclonal	microtiterplate	0.08 $\mu g\ l^{-1}$
	J. T. Baker B.V., Groß Gerau, Germany	polyclonal	magnetic particles	0,01 $\mu g\ l^{-1}$
Carbofuran	AVENA GmbH Markt Schwaben, Germany	polyclonal	tubes	0.05 $\mu g\ l^{-1}$
	J. T. Baker B.V. Groß Gerau, Germany	polyclonal	magnetic particles	0.06 $\mu g\ l^{-1}$
	Immuno Systems, Inc., Scarborough, ME, USA	polyclonal	tubes	1 $\mu g\ l^{-1}$
Chlordane	AVENA GmbH Markt Schwaben, Germany	polyclonal	tubes	0.5 $\mu g\ l^{-1}$
Chlorpyriphos	J. T. Baker. Groß-Gerau, Germany	polyclonal	magnetic particles	0,1 $\mu g\ l^{-1}$

Table 2. (continued)

Analyte	Producer/Distributor	Antiserum	Assay format	Detection limit*
Cyanazin	J. T. Baker. Groß-Gerau, Germany	polyclonal	magnetic particles	0.04 µg l^{-1}
2.4-D	AVENA GmbH Markt Schwaben, Germany	polyclonal	microtiterplate	1.0 µg l^{-1}
	J. T. Baker B.V. Groß-Gerau, Germany	polyclonal	magnetic particles	0,7 µg l^{-1}
	Immuno Systems, Inc., Scarborough, ME, USA	polyclonal	tubes	1.0 µg l^{-1}
Metalaxyl	AVENA GmbH Markt Schwaben, Germany	polyclonal	microtiterplate	0.1 µg l^{-1}
	Immuno Systems,Inc., Scarborough, ME, USA	polyclonal	tubes	0.5 µg l^{-1}
Methoprene	AVENA GmbH Markt Schwaben, Germany	polyclonal	microtiterplate	1.0 µg l^{-1}
Metolachlor	J. T. Baker B.V. Groß-Gerau, Germany	polyclonal	magnetic particles	0.05 µg l^{-1}
Pentachloro-phenol	Westinghouse Bio-Analytic Systems, Rockville, Maryland, USA	polyclonal	microtiter plates	
Triazines (Atrazine Simazine, Propazine)	Transia GmbH Obermörlen, Germany	polyclonal	tubes	0,1 µg l^{-1}
Triazines	AVENA GmbH Markt Schwaben, Germany	polyclonal	microtiterplate tubes	0.02 µg l^{-1} 0.1 µg l^{-1}
	Millipore GmbH Eschborn, Germany	polyclonal	microtiterplate tubes	0.02 µg l^{-1} 0.1 µg l^{-1}
	R-Biopharm GmbH Darmstadt, Germany	polyclonal	microtiterplate	0.01 µg l^{-1}
	Immuno Systems Inc., Scarborough, ME, USA	polyclonal	tubes	1 µg l^{-1}

*Declaration of the Producers/Distributors

lend itself to a commercial test, in particular because of the difficulties involved with reagent disposal, and the danger of damage to human health by handling of radioactive substances. Following the development of lower cost photometers for the rapid and simultaneous measurement of extinctions, the polystyrol microtitre plate has gained increasing acceptance, since it permits simultaneous production of the standard curve with a large enough number of single values obtained under almost identical measurement conditions.

Some manufacturers offer their developments in tubes or use cards to build up the heterogeneous assays. Not all of the tests listed in Table 2 were originally developed for drinking, surface, and ground waters, but they could be applied to the water matrix without difficulty, on condition that the detection limit allows a direct measurement.

One should not fail to emphasise at this point that drinking and surface waters, according to location and catchment area, present in every way an unpleasant, difficult object for examination. The humic substances especially, being pretty well ubiquitous in many locations, hinder the immunochemical measurement by the appearance of false positive results in ways which are mainly unclear at the present. Solid phase or liquid-liquid extractions for analyte enrichment or separation from interfering matrix components, preceding the immunochemical test, does not always produce the desired result. One should also consider that the substantial advantages of the immunoassay technique, such as procedural rapidity, high sample throughput, and reduction of solvent consumption must be lost in such a case.

A not inconsiderable number of experts in the area of water analysis hold most commercial assays to be, at best, semi-quantitative. One may not agree with this opinion in every case, it is however clear that the unlimited and uncritical adoption of such assays can lead to serious difficulties when carrying out the analysis. Requirements for home made assays are also of course valid to an even greater extent for commercial assays. Should they be used as screening methods, then positive results must be adequately confirmed in each case.

According to users' experiences, appropriate care should be taken with manufacturers' statements concerning their so called "group specific assays", equally suited for the measurement of a large number of chemical substances of similar character with more or less the same sensitivity. The development of "group specific assays", based on polyclonal antibodies, with which, in addition to the true analyte (cross reactivity 100%), all compounds in the group may be detected with almost the same sensitivity (detection limit) is extremely difficult, if not principally impossible.

The production of group specific assays based on a mixture of monoclonal antibodies may bring a solution to the problem. Hock, together with the study group [95] carried out measurements of s-triazines in water samples with antibodies, respectively test kits from a number of manufacturers. The results confirmed, in principle, the suitability of immunoassays for the detection of s-triazines. The exact description of the assay, including standard curve, detection limit, cross reactivities and the necessary information about the possible presence of matrix effects is, according to the opinion of experimental workers, a condition for successful application of the immunochemical method. Careful drawing up of the calibration curve is required, along with working in the test's middle region. With a sample matrix of unknown composition, the sample should be spiked using known standards.

2.7 Confirmation and Validation

Considering the development of immunoassays for pesticide determination, one can only come to the conclusion that, just as with biosensors, a large number of home made assays exists in comparison to the relatively small number of commercially produced assays. This is especially the case for those assays suitable for the examination of ground and surface waters. Many methods of this type have only been developed for use in basic research. Often, because of the most various shortcomings, they cannot achieve the status of being established, practicable methods. It may be assumed that the actual number of immunoassays is far greater than indicated by the relevant literature. Pesticide manufacturers and university research groups working in the most varied fields possess assays, or at least the appropriate antisera, without scientifically publishing in every case. The discrepancy between the relatively large number of new developments for research purposes over the last ten years and the very limited number of assays usable in practice has certainly many grounds. Having contact with, and dealing with biochemically based measuring principles is unfamiliar territory for many of the more traditionally educated chemists, analysts, and environmental scientists. The terminology is often difficult to grasp, and the use of the relevant, occasionally inconsistent, nomenclature not without its difficulties. On the other hand, the simplicity of carrying out these techniques is emphasised by manufacturers of commercial assays. A thorough training of specialists appears to be seen to be unimportant.

The main reason for the hesitant development of the immunochemical methodology in the field of pesticide analysis lies in the special characteristics of the analytic principle itself, in which the matrix complexity naturally plays a large part.

The high affinity constant of the antigen/antibody reaction determines the low detection limits, a consequence of which is that water samples can be examined, in general, without preparative measures (enrichment, derivatisation etc.). This must be noted when interpreting the detection limits given in Table 2. The microtitre plate format permits a high sample throughput, meaning altogether short analysis times, and, because of the high number of measurements, reliable information about their precision.

The second excellent characteristic of polyclonal and monoclonal antibodies is their distinct selectivity (specificity). High selectivity does not, however, mean absolute selectivity. Cross-reactions between the antibodies and compounds with similar structures to the antigens are responsible for the fact that, to a limited extent, some samples will be recorded as positive although they contain no actual analyte (false positive samples). On the other hand, false negatives are rare with a well optimised assay, but not impossible. The certainty of any statement which relates to the exceeding of legally prescribed limits depends on the difference between the detection limit and the legal limit. In spite of high affinity and selectivity, the enzyme immunochemical assay possesses one critical disadvantage relative to the standard chromatographic-spectrometric techniques. The signals arising as response to the concentration are not time resolved and no further resolution of the energy content of the signal results from these methods. Each signal (value) is practically a sum from all cross reacting components present in the solution being measured. Immunoassays are not suited to the examination of multi-component mixtures, but are fine for the determination of single substances. Even here, ideas of the matrix composition must, of course, exist. One must at least be certain that cross reactions will not occur to any great extent. When cross reactions are present to a greater degree, then the presence and concentration of the actual analyte can only be correlated to the measured value in special cases. Without doubt, the greatest importance of immunoassays is therefore as a screening method.

It is not possible here to go into the details of any single piece of work over the problematics of confirmation. GC/MS is often raised to the level of absolute confirmation method. Likewise for less volatile analytes the system LC/MS.

Summarising, it must be noted that a very good correspondence for standard deviations and recovery quotas have been reported in publications relevant to this theme. The less well developed selectivity of immunoassays relative to the GC/MS combination is expressed in the larger number of positive results which must be accepted, as long as they do not exceed a specified limit, and the number of false negative samples remains small.

From this it is conversely clear that for assay optimisation, tests and details concerning the extent of possible cross reactions are imperative. It must also be clear that the application of immunochemical methods must always be supported by validation and confirmation. The value of carrying out immunochemical tests without confirmation by chromatographic-spectrometric techniques would seem questionable.

3 Enzyme Inhibition

Use is made of the inhibiting effect of particular pesticides on enzyme activity in the enzyme inhibition method. These methods – originating mainly from the period between 1965 and 1975 – have retained some importance because they are relatively simple and of reasonable cost. Of the older methods of this type, the determination of organophosphorus compounds and carbamates in surface water based on the inhibition of cholinesterase has maintained a certain importance.

Enzyme inhibition techniques are not highly selective. When several inhibiting pesticides are present, only a qualitative statement may be made due to the different inhibiting activities of the single components. There was no lack of attempts to increase the little developed selectivity of this method by preceding chromatographic separation steps. An initial separation by thin layer chromatography is described by Horvath (135). Ramsteiner and Hörmann (136) report a separation of organophosphorus compounds and carbamates by HPLC, and subsequent detection by cholinesterase inhibition. The toxic nature of organophosphorus compounds is traceable to their ability to inhibit cholinesterase. Even though their ranges of application are limited, two techniques still have a certain importance, and are briefly exemplified here.

3.1 Radiometric Method (135)

The analyte inhibits the enzyme cholinesterase, acetylcholine acetylhydrolase, EC 3.1.1.7:

$$[^3\text{H}]\text{acetylcholine} + \text{H}_2\text{O} \xrightarrow{\downarrow\ \text{inhibitor ChE}} [^3\text{H}]\ \text{acetate} + \text{choline}$$

The acetic acid produced by the reaction is shaken with the liquid scintillation cocktail. The aqueous phase contains the remaining [^{3}H]acetylcholine which has not been converted. Beta particles emitted from the [^{3}H]acetylcholine which has not been hydrolysed do not reach the scintillation solution owing to their low energy. Thio- or dithiophosphates can be oxidised with bromine to their "oxon" analogues, which are generally much stronger cholinesterase inhibitors than the original compounds.

With respect to the count rate measurement in the scintillation spectrometer, a relative standard deviation of +/ − 0.68% has been reported.

The detection limits for pesticides are spread over a wide range: parathion 8.88×10^{-7} mol l^{-1}, paraoxon 4.29×10^{-10} mol l^{-1}, malathion 2.19×10^{-7} mol l^{-1}, and dichlorvos 1.28×10^{-7} mol l^{-1}. The method is held to be pretty sensitive, but not selective.

3.2 Colorimetric Method

Indophenyl acetate, (n-4-acetoxyphenyl)-4-quinone, may be oxidised using cholinesterase to an intensively blue coloured compound. The rate of hydrolysis of the indophenyl acetate is reduced by cholinesterase inhibiting pesticides. The change in absorbance at 650 nm per unit time, compared with the non-inhibited reaction, is a measure of the amount of pesticide residue present:

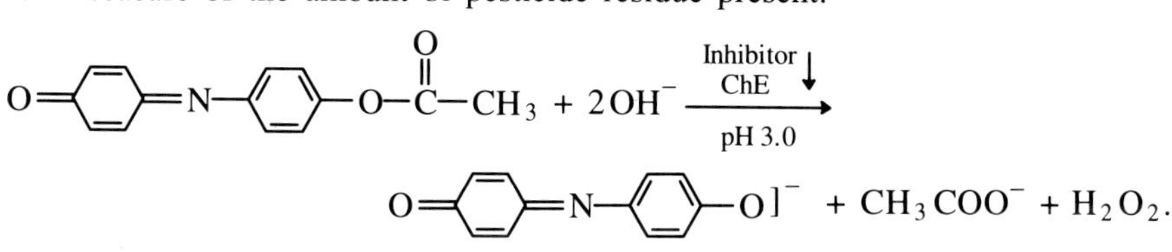

The relative standard deviation at a residue level of 0.1 mg/l for dichlorvos lies at 3.7%. The detection limit lies at 0.01 mg/l water.

Summarising, it must be said that the enzyme inhibition methods have to a great extent lost their importance as screening techniques in water analysis due to their poor or failing selectivities. Also, when considering detection limits (sensitivity), the inhibition assays are inferior to the classic type of immunoassay, and most definitely to the standard chromatographic-spectrometric methods.

4 Biosensors

The term 'biosensors' is taken to mean those measuring systems in which selectively binding and reacting biomolecules are closely coupled with a so called 'transducer'. The physicochemical signal resulting from the reaction between receptor component and analyte is converted into an electrical signal in the transducer and finally subjected to electronic data processing (Fig. 6). Suitable receptor components are micro-organisms, enzymes, enzyme systems, cell tissue, organelles and antibodies. On the transducer side are mainly electrochemical (e.g. amperometric oxygen, potentiometric pH and ammonium electrodes), optical, and optoelectronic measuring methods, which, combined with the receptors, take part in the signal production.

A great demand exists for the development of biosensors. Biosensors possess in their application, relative to traditional chromatographic-spectrometric methods, in the first instance, the same advantages as those from the development of the older immuno- and enzyme assays. High sample frequency, extensive avoidance of environmentally harmful solvents and chemicals through avoidance of sample preparation, and simple analytical procedure facilitated by the mobility of the measuring apparatus, allowing in-situ measurement, speak for the biosensor.

The selectivity of the biosensor is specified by the choice of the suitable receptor. With a matrix of unknown composition – and especially for residue analysis of drinking and waste water – the danger of interferences appearing is always present. For this reason care must be taken to watch for cross reactions on the receptor. Biosensors, excepting immunosensors, are distinguished negatively

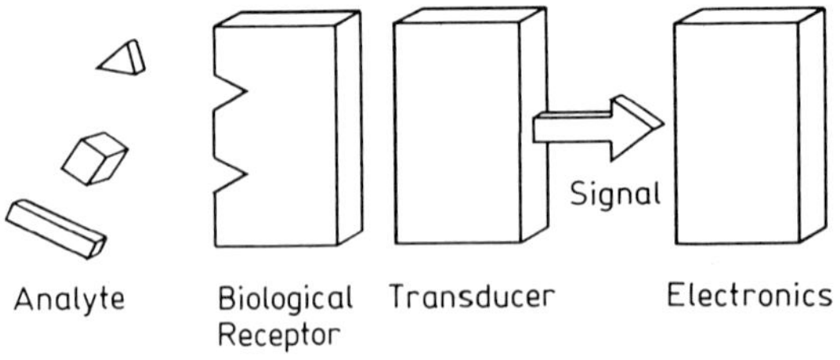

Fig. 6. The function of a biosensor

by their high susceptibility to toxically active substances (enzyme toxins). Receptors of biological origin possess only a limited shelf life. They especially tend to denature irreversibly at the binding sites to the carriers and membranes, and to lose selective binding character. It is conspicuous that, although the number of relevant publications has shown a rapid development since the description of the first biosensor by Clark (137) in 1962, this has not precipitated a larger availability of really fully developed commercial biosensors. According to estimates from Renneberg and Gründig (138) up to the present, circa 120 biosensors for the determination of diverse analytes have been described, only 10% of which are commercially available sensors.

A number of biosensor measuring principles have been described for the special problematics of pesticide determination in drinking and ground water. They are mainly enzyme inhibition sensors which contain cyanobacteria, thylakoid membranes, protoplasts or also immobilised enzymes (acetyl-butyryl-cholinesterase) as receptor components. In general, enzymes, enzyme systems, or the electron transport system of intact cells are blocked by the analyte. The rate of the enzyme catalysed reaction sinks to the same extent, and hence the measured signal, which thus reflects the 'toxicity' of the substances in the sample relative to the receptor as integral sum.

The cholinesterase sensor (139) is still the best known and most often described sensor for the determination of organophosphate pesticides and carbamates. Butyrylthiocholine as substrate is hydrolysed by the cholinesterase immobilised on the sensor head to thiocholine, which is easily oxidised on a platinum electrode. The lower the concentration of inhibited analyte, the higher is the remaining activity of the partially inhibited enzyme. The detection limits of inhibitor sensors based on cyanobacteria lie around 20 $\mu g\,L^{-1}$ for the herbicides atrazine, metoxuron, and linuron and are thus inferior to the respective immunoassays. Similarly, the determination of the herbicides bentazon (142, 143), deiquat and diuron with detection limit of 2–4 $\mu g\,L^{-1}$ using a sensor based on broad bean receptors does not fulfil all demands. For the determination of the bromophos-ethyl, a detection limit of 10 $\mu g\,L^{-1}$ has been reported (139). The corresponding values for paraoxon and malathion lie by 2 $\mu g\,L^{-1}$ and 6 $\mu g\,L^{-1}$, respectively, well over the limits for the comparable immunoassays.

Although the use of antibodies as receptor components, not least because of the great success of immunoassays, presses onward, their coupling with a powerful transducer is not without its problems. It is certainly correct that the employment of immunoglobulin brings to the sensor excellent qualities as regards the selectivity and sensitivity. On the other hand, the direct measurement of antigen binding by optoelectronic means produces relatively weak signal intensities. It is therefore usual, in the main, to prefer indirect measurement of the antigen-antibody reaction by the addition of tracers, as with the enzyme immunoassays. By using hapten-glucoseoxidase conjugates as competitive reagent for the free analyte, hydrogen peroxide will be generated from glucose and oxygen in an amount corresponding to that of the free enzyme. This amount is the greater, the lower was the original content of free hapten (analyte). An interesting further development in the immunosensor field is described by Renneberg and Gründig (138). Since achieving equilibrium in immunosensors requires some time (measuring time 10 to 20 min), the authors recommend as basic building block for an alarm sensor to measure atrazine in water samples a displacement sensor. Immobilised antibodies are pre-incubated with antigen (hapten) enzyme markers of relatively lower affinity. By addition of the sample with high affinity antigens, the enzyme tracers are freed from the antibodies, transported by the elution flow, converted with substrate, and the resulting product is measured electrochemically.

The application of reflectometric interference spectrometry, with which one can measure the strength of extremely thin layers, opened-up, according to Gauglitz et al. (140) the possibility of following the binding process between antigen and antibody in real time. The condition, that small molecules to be detected only make a minimal difference to the refraction index by their coating the sensor is taken into account in that, similarly to an immunoassay, a tracer consisting of the analyte and a further high molecular weight component is added at a known concentration to the measuring solution. In the case of a pesticide analysis in water, the relatively small analyte molecules compete with the very large hapten-protein conjugates for the binding sites of the excess of antibody present and produce thicker and thus more easily measured layers.

Further biosensor development depends on a great number of factors. Improvements over the, at present, mainly non-commercial biosensors are on both the receptor and transducer sides both feasible and necessary. The application, respectively binding, of the receptors in thin, firmly bound and durable layers can not be counted as completely solved. Considering the practicability of a biosensor, sensitivity (detection limit) and selectivity (specificity) are not the only factors of importance. Most of all the response and the reversibility control the value of the sensor for the practician. In this area, too, decisive improvements are required, which can only be achieved by the co-operation of biotechnologists, analysts, and experts in the fields of microelectronics and instrumentation. The commercial potential of the biosensors is heavily dependent on the applicability. In spite of all the potential advantages over usual chromatographic-spectrometric techniques, a biosensor will never achieve the selectivity and diagnostic value of GC/MS. The value of biosensor measurement lies, just as for the meanwhile classic immuno- and enzyme assays, in the reduction of false negative test results. It is evident, that biosensors and usual high resolution techniques complement each other excellently. The high requirements of consumer and legislature for drinking and ground water quality will certainly ensure that the biochemically based sensors' development will remain a stimulating and interesting area on the borders between chemical analysis and microelectronics.

5 Conclusion and Future Developments

When one looks at the developments from the last 10–15 years, one can summarise by saying that the biochemical methods have still to be accepted to more than a very limited extent, this being equally valid for the immunoassays, the enzyme inhibition tests and the biochemically based sensors. The most likely case is that of the immunoassay, which presents, with its excellent sensitivity, a valuable supplement to the usual chromatographic-spectrometric methods. The future area of application of immunoassays in the field of residue analysis with drinking and ground water matrices doubtless lies in their use as screening methods. Use of the immunochemical techniques without basic confirmation of their results presents risks in both qualitative and quantitative aspects. This does not preclude, when the matrix composition is known, that based on the high sample frequency and simple apparatus, such methods may be suitable for 'in the field' analyses. A likely possibility for development in the coming years lies in the immunoaffinity chromatography. Immunoaffinity chromatography enables a united clean up and enrichment, and presents a welcome addition to the standard methods. Ultimately, affinity chromatography opens up the possibility of developing group-selective determinations, which allow the measurement of a greater number of pesticides of similar chemical structures in the water matrix. Further future development of enzyme inhibition tests is not to be expected. This methodology is, as a rule, inferior to the immunochemical methods both in sensitivity and selectivity. This holds also for the biosensors, as long as they function on the basis of micro-organisms, organelles and enzymes as receptors. Without considering the prospective transducer developments, it would also appear that here the immunosensors with antibodies as receptors have the best chances for development. Last but not least, it must be brought to the readers' attention that the technical development must be accompanied by some officially organised efforts to recognise the biochemical techniques as validated, practicable routine methods. In this context, the preparation and supply of certified materials plays a decisive role.

Acknowledgement. I am extremely grateful to Mr. R.D. Hatton for the translation of my contribution.

6 References

1. Van Emon JM, Seiber JN, Hammock BD (1989) Immunoassay techniques for pesticide analysis. In: Analytical methods for pesticides and plant growth regulators, vol XVII Academic Press, p 217
2. O'Sullivan MJ (1984) Enzyme immunoassay. In: Butt WR (ed) Practical immunoassay, The state of the art. Marcel Dekker, New York, p 37
3. Schwalbe-Fehl M (1989) Immunoassays für niedermolekulare umweltrelevante Verbindungen. In: Borsdorf R, Fresenius W, Günzler H, Huber W, Kelker H,

Lüderwald I, Tölg G, Wisser H, (eds) Analytiker-Taschenbuch, Bd. 8, Springer, Berlin Heidelberg New York, p 247
4. Chan DW (1987) General principle of immunoassay. In: Chan DW (ed), Immunoassay a practical guide Academic Press, Inc. San Diego, p 1
5. Tijssen P (1985) Burdon RH and van Knippenberg PH (eds) Practice and theory of enzyme immunoassay, Amsterdam, p 221
6. Kabakoff DS (1980) In: Maggio ET (ed) Enzyme immunoassay, CRC Press, Boca Raton, p 71
7. Bauminger S, Wilchek M (1980) In: Langone JJ, Van Vunakis H (eds) Methods in enzymologie, Immunochemical techniques, vol. 70 Part A., Academic Press, New York, p 151
8. Bodanszky M, Bodanszky A (1984) The practice of peptide syntheisis. Springer, Berlin Heidelberg, New York, p 87
9. Knoll E (1989) Radioimmunoassay In: Borsdorf R, Fresenius W, Günzler H, Huber W, Kelker H, Lüderwald I, Tölg G, Wisser H (eds) Analytiker-Taschenbuch, Bd. 8, Springer, Berlin Heidelberg, New York, p 93
10. Vaughan JR, Osato RL (1951) J. Am. Chem. Soc. 73: 5553
11. Erlanger BF (1980) In: Van Vunakis H, Langone JJ, (eds) Methods in enzymology vol. 70, Academic Press, New York, p 85
12. Goodfriend TL,Levine L, Fasman GD (1964) Science (Washington) 144: 1344
13. Hung BT, Benner SA, Williams CM (1980) J. Biol. Chem. 255: 6047
14. Tijssen P (1985) Burdon RH and van Knippenberg PH (eds), Practice and theory of enzyme immunoassay, Amsterdam p 243
15. Scatchard G (1949) Ann. N.Y. Acad. Sci. 51: 660
16. Köhler G, Milstein C (1975) Nature (London) 256: 495
17. Langone JJ, Bjerke RJ (1989) Anal. Biochem. 182: 187
18. Huse WD, Sastry L, Iverson SA, Kang AS, Alting-Mees M, Burton DR, Benkovic SJ, Lerner RA (1989) Science, 246: 1275
19. Ward SE, Güssow D, Griffiths AD, Jones PT, Winter G (1989) Nature 341: 544
20. Berson SA, Yalow RS (1968) Clin. Chim. Acta 22: 51
21. Engvall E, Perlmann P (1971) Immunochemistry 8: 871
22. Van Weeman BK, Schuurs AHWM (1971), FEBS Lett. 15: 232–235
23. Hock B (1988) Bestimmung von Pflanzenschutzmitteln mit Hilfe von Enzymimmunoassays 12. Aachener-Werkstattgespräch vom 15.9.–16.9.1988
24. Walker WHC (1977) Clin. Chim. 23: 384–402
25. Ercegovich CD (1971) Analysis of pesticide residues: immunological techniques, In: Biros FJ (ed) Pesticide identification at the residue level, Advances in Chemistry Series No. 104, American Chemical Society, Washington DC, p 162
26. Centeno ER, Johnson WJ, Sehon AH (1970) Int. Arch. Allergy Appl. Immunol. 37: 1
27. Wratten SJ, Feng PCC (1990) Pesticide analysis by immunoassay In: Rittenburg JH, Development and application of immunoassay for food analysis Part III, Chapter 8. Elsevier Applied Science, London
28. Van Emon JM (1992) Anal. Chem. 64: 79
29. Kaufman BM, Clower M (1991) J. Assoc. Off. Anal. Chem. 74: 239
30. Van Emon JM, Seiber JN, Hammock BD (1989) Immunoassay techniques for pesticide analysis In: Sherman (ed) Analytical methods and plant growth regulators, Academic Press, San Diego
31. Jung F, Gee SJ, Harrison RO, Goodrow MH, Karu AE, Braun AL, Li QX, Hammock BD (1989) Use of immunochemical techniques for the analysis of pesticides. Pestic. Sci. 26: 303–317
32. Van Emon J (1989) Abstr. 46, 198th National Meeting, American Chemical Society, Miami Beach, FL, September 1989
33. Anonymous (1989) J. Assoc. Off. Anal. Chem. 72: 694–704

34. Mastrorocco D, Brodsky M (1990) J. Assoc. Off. Anal. Chem. 73: 331–332
35. Anonymous (1989) Fed. Regist. 54, 33920–33923
36. Lukens HR, Williams CB, Levinson SA, Dandliker WB, Muragama D, Baron RL (1977) Environ. Sci. Technol. 11: 292
37. Newsome WH, Shields JB (1981) J. Agric. Food Chem. 29: 220
38. Newsome WH, Collins PC (1987) J. Assoc. Off. Anal. Chem. 70: 1025
39. Jung F (1988) Dissertation, Univ. Stuttgart-Hohenheim, Germany
40. Jung F, Meyer HHD, Hamm RT (1989) J. Agric. Food Chem. 37: 1183
41. Newsome WH (1987) In: Greenlaich R, Roberts TR (eds) Pesticide science and biotechnology, Blackwell Scientific Publications, London, p 349
42. Deschamps RJA, Wilson L, Hall JC (1990) Seventh Int. Congress of Pesticide Chemistry vol. III Poster 08E - 20, Hamburg, Aug. 5–10, p 352
43. Newsome WH (1985) J. Agric. Food Chem. 33: 528
44. Klausner A (1987) Bio-Technology 5: 551
45. Sharp CR, Feng PCC, Horton SR, Logusch EW (1989) 198th ACS National Meeting, Division of Agrochemicals, Miami Beach, Florida, paper 30
46. Feng PCC, Wratten SJ, Horten SR, Sharp CR, Logusch EW (1990) J. Agric. Food Chem. 38: 159
47. Sharp CR, Feng PCC, Horton SR, Logusch EW (1991) Development of highly specific antibodies to alachlor by use of a carboxy-alachlor protein conjugate In: Tweedy et al. (eds) Pesticide residues and food safety – a harvest of view points. ACS Symposium Series 446. Am. Chem. Soc., Washington DC, p 87
48. Huber SJ (1985) Chemosphere 14: 1795
49. Jung F, Szēkācs A, Li Q, Hammock BD (1991) J. Agric. Food Chem. 39: 129
50. Schlaeppi JM, Förg W, Ramsteiner K (1989) J. Agric. Food Chem. 37: 1532
51. Wittmann C, Hock B (1990) Food and Agric. Immunology 2: 65
52. Wittmann C, Hock B (1990) Seventh Int. Congress of Pesticide Chemistry, vol. III, Poster O8E-09, Hamburg, Aug. 5–10, p 341
53. Wüst S, Hock B (1990) Seventh Int. Congress of Pesticide Chemistry, vol. III, Poster O8E-10, Hamburg, Aug. 5–10, p 342
54. Lucas AD, Schneider P, Harrison RO, Seiber JN, Hammock BD, Biggar JW, Rolston DE (1991) Food and Agric. Immunol. 3: 155
55. Wittmann C, Hock B (1991) J. Agric. Food. Chem. 39: 1194
56. Bushway RJ, Perkins B, Savaga SA, Lekousi SJ, Ferguson BS (1989) Toxicol. 42: 899
57. Bushway RJ, Perkins B, Savaga SA, Lekousi SJ, Ferguson BS (1988) Bull. Environ. Contam. Toxicol. 40: 647
58. Schneider P, Hammock BD (1992) J. Agric. Food Chem. 40: 525
59. Rubio FM, Itak JA, Scutellaro AM, Selisker MY, Herzog DP (1991) Food and Agric. Immunology 3: 113
60. Weil L, Schneider RJ, Schäfer O, Ulrich P, Weller M, Ruppert T, Niessner R (1991) Fresenius J. Anal. Chem. 339: 468
61. Wüst S, Doht U, Giersch T, Wittmann C, Hock B (1990) GIT Fachz. Lab. 34: 99
62. Wittmann C, Hock B (1989) Food and Agric. Immunol. 1: 211
63. Wartberg M, Cammann K, Strupat K, Hillenkamp F, (1994) Fresenius J. Anal. Chem. 348: 240
64. Giersch T, Hock B (1990) Food and Agric. Immunol. 2: 85
65. Li QX, Hammock BD, Seiber JN (1991) J. Agric. Food Chem. 39: 1537
66. Szurdoki F, Bekheit HKM, Marco M-P, Goodrow MH, Hammock BC (1992) J. Agric. Food Chem. 40: 1459
67. Reck B, Gerhard JUE, Frevert J, Knoell HE (1990) Seventh Int. Congress of Pesticide Chemistry, vol. III Poster O8E-05, Hamburg, Aug. 5–10, p 337
68. Krotzky AJ, Candels U, Schweda Z (1990) Seventh Int. Congress of Pesticide Chemistry, vol. III, Poster O8E-12, Hamburg, Aug. 5–10, p 334

69. Kelley MM, Zahnow EW, Peterson WC, Toy ST (1985) J. Agric. Food Chem. 33: 962
70. Rinder DF, Fleeker JR (1981) Bull. Environ. Contam. Toxicol. 26: 375
71. Fleeker J (1987) J. Assoc. Off. Anal. Chem. 70: 874
72. Newsome WH, Collins PG (1989) Food and Agric. Immunol. 1: 203
73. Knopp D, Nuhn P, Dobber KANHJ (1985) Arch. Toxicol. 58: 27
74. Weber W, Rubach K (1994) Acta. Hydrochim. Hydrobiol. 2: 147
75. Hall JC, Deschamps RJA, Krieg KK (1989) J. Agric. Food Chem. 37: 981
76. Wittmann C, Hock B (1991) J. Agric. Food Chem. 39: 1194
77. Wittmann C, Hock B (1991) Z. Wasser-Abwasser-Forsch. 24: 2
78. Wittmann C, Hock B (1990) Vom Wasser 75: 115
79. Schwalbe M, Dorn E, Beyermann K (1984) J. Agric. Food Chem. 32: 734
80. Newsome WH, Collins PG (1990) Food and Agric. Immunol. 3: 127
81. Gee SJ, Harrison RO, Goodrow MH, Braun AL, Hammock BD (1991) Development of immunoassays for thiocarbamate herbicides In: Vanderlann M, Stanker LH, Watkins BE, Roberts DW (1991) (eds) Immunoassays for trace chemical analysis – monitoring toxic chemicals in humans, food and the environment. ACS Symposium Series 451, Am. Chem. Soc. Washington DC (1991) p 100
82. Poppe F (1994), Dissertation D 83, Technische Universität Berlin
83. Scholz H, Hock B (1990) Seventh Int. Congress of Pesticide Chemistry, vol. III, Poster O8E-07, Hamburg, Aug. 5–10, p 339
84. Scholz H, Hock B (1991) Z. Wasser-Abwasser-Forsch. 24: 29
85. Kreißig S, Hock B (1990) Seventh Int. Congress of Pesticide Chemistry, vol. III, Poster, Hamburg, Aug. 5–10
86. Kreißig S, Hock B (1991) Z. Wasser-Abwasser-Forsch. 24: 10
87. Schlaeppi JM, Moser H, Ramsteiner K (1991) J. Agric. Food Chem. 39: 1533
88. Li QX, Gee SJ, Mc Chesney MM, Hammock BD, Seiber JN (1989) Anal. Chem. 61: 819
89. Harrison RO, Braun AL, Gee SJ, O'Brien DJ, Hammock BD (1989) Food and Agric. Immunol. 1: 37
90. Gee SJ, Miyamoto T, Goodrow MH, Buster D, Hammock BD, (1988) J. Agric. Food Chem. 36: 863
91. Gee SJ, Miyamoto T, Goodrow MH, Buster D, Hammock BD (1988) J. Agric. Food Chem. 36: 863
92. Sidki AM, Al-Abdulla IH, Rowell FJ (1987) Clin. Chim. 33: 463
93. Riggle BD, Dunbar B (1987) Abstr. Meet. Weed Sci. Sco. Am. 27: 73
94. Riggle BD, Dunbar B (1990) J. Agric. Food Chem. 38: 1922
95. Van Emon J, Seiber J, Hammock B (1987) Bull. Environ. Contam. Toxicol. 39: 490
96. Niewola Z, Benner JP, Swaine H (1986) Analyst 111: 399
97. Levitt T (1972) Proc. Anal. Div. Chem. Soc. 16: 72
98. Levitt T (1977) Lancet 8033: 358
99. Fatori D, Hunter WM (1980) Clin. Chim. Acta 100: 81
100. Niewola Z, Walsh ST, Davies GE (1983) Int. J. Immunopharmac. 5: 211
101. Van Emon JM, Seiber JN, Hammock BD (1985) ACS Symp. Ser. 276: 307
102. Van Emon J, Hammock BD, Seiber J (1986) Anal. Chem. 58: 1866
103. Hall JC, Deschamps RJA, Krieg KK (1989) J. Agric. Food Chem. 37: 981
104. Deschamps RJA, Hall JC (1991) Food and Agric. Immunol. 3: 127
105. Deschamps RJA, Hall JC, Mc Dermott MR (1990) J. Agric. Food Chem. 38: 1881
106. Giersch T (1991) Z. Wasser-Abwasser-Forsch. 24: 7
107. Kuniyuki KH, Mc Carthy S (1986) Abstr. of papers, The 6th Int. Congress of Pesticide Chemistry, Ottawa, Canada, INPAC 1986, Abstr. 5C-11
108. Huber SJ, Hock B (1985) Z. Pfl. Krankh. Pfl.-Schutz 92: 147
109. Huber SJ, Hock B (1985) GIT Fachz. Lab. 29: 969

110. Newsome WH (1986) Bull. Environ. Contam. Toxicol. 36: 9
111. Wittmann CH, Hock B (1989) Food and Agric. Immunol. 1: 211
112. Giersch T, Hock B (1990) Food and Agric. Immunol. 2: 85
113. Hock B (1989) Z. Wasser-Abwasser-Forsch. 22: 78
114. Huber SJ, Hock B (1986) In: Bergmeyer U (ed), Methods of enzymatic analysis, VCH Weinheim, vol. XII, p 438
115. Dunbar B Riggle B, Niswender G (1990) J. Agric. Food Chem. 38: 433
116. Wüst S, Doht U, Giersch TH, Wittmann CH, Hock B (1990) GIT Fachz. Lab 2: 99
117. Hock B, together with Immunoassay Study Group (1991) Anal. Letters 24: 529
118. Harrison RO, Goodrow MH, Hammock BD (1991) J. Agric. Food Chem. 39: 122
119. Weil L, Schneider J, Schäfer O, Ulrich P, Weller M, Ruppert T, Neißner R (1991) Fresenius J. Anal. Chem. 339: 468
120. Fleeker R, Mumma RO (1987) Abstr. Pap. Am. Chem. Soc. 194 Meet, Agro 146
121. Langone JJ, Van Vunakis H (1975) Res. Commun. Chem. Pathol. Pharmacol. 10: 163
122. Wie SJ, Hammock BD (1982) J. Agric. Food Chem. 30: 949
123. Wing KD, Hammock BD (1979) Experientia 35: 1619
124. Wing KD, Hammock BD, Wustner DA (1978) J. Agric. Food Chem. 26: 1328
125. Reck B, Harperath U, Ringel J, Frevert J (1990) Seventh Int. Congress of Pesticide Chem. vol. III Poster O8E-09, Hamburg, Aug. 5–10, p 338
126. Wraith MJ, Hitchings EJ, Woodbridge AP, Cole ER, Roberts TR (1986) Sixth Int. Congress of Pesticide Chem., Poster Session
127. Centeno ER, Johnson WJ, Schon AH (1970) Int. Arch. Allergy 37: 1
128. Wie SJ, Hammock BD (1984) J. Agric. Food Chem. 32: 1294
129. Wie SJ, Sylwester AP, Wing KD, Hammock BD (1982) J. Agric. Food Chem. 30: 943
130. Vellejo RP, Bogus ER, Mumma RO (1982) J. Agric. Food Chem. 30: 572
131. Ercegovich CD, Vallejo RP, Gettig RR, Woods L, Bogus ER, Mumma RO (1981) J. Agric. Food Chem. 29: 559
132. Hunter KW, Lenz DE (1982) Life Sci. 30: 355
133. Wengatz J, Schmid RD (1990) Seventh Int. Congress of Pesticide Chem. vol. III, Poster O8E–08, Hamburg, Aug. 5–10, p 340
134. Cook CE, Ballentine NH, Seltzman TB, Fallent CR (1979) J. Pharmacol. Exp. Therap. 210: 391
135. Horrath L, Forster T (1980) Radiometric enzymic method for determining organophosphorus and carbamate residues In: agrichemica residue – biota interactions in soil and aquatic ecosystems, IAEA STI/PUB/548, Vienna
136. Ramsteiner KA, Hörmann WD (1975) J. Chromatogr. 104: 438
137. Clark LC, Lyons C (1962) Ann. NY Acad. Sci. 102: 29
138. Renneberg R, Gründig B (1994) Nachr. Chem. Techn. Lab. 1: 42
139. Campanella L, Cocco R, Sammeartino MP, Tomassetti M (1992) Sci. Total. Environ. 123/124: 1
140. Gauglitz G, Brecht A, Ingenhoff J, Kraus G (1994) Spektr. der Wissenschaft 1: 92
141. Weiler EW (1990) In: Haug G et al. (ed) Chemistry of plant protection. vol. 3: Pyrethroid residues, immunoassays for low molecular weight compounds, 147–228
142. Lindner SE, Obermayer S, Schnablh (1992) Angew. Bot. 66: 79
143. Obermayer S′, Lindner SE, Hostert E, Schnabl H (1993): Der protoplasten biotest – ein Wirkungstest zur Summarischen Schadstofferfassung in der Umwelt. Abstr. Bio Test-Symp. des Bundesgesundheitsamtes, Berlin 10.–14. Mai (1993)

Growth Inhibition of Plants as a Bioassay for Herbicide Analysis

Wilfried Pestemer and Petra Günther

Federal Biological Research Centre for Agriculture & Forestry, Institute for Ecological Chemistry, Königin-Luise-Str. 19, D-14195 Berlin-Dahlem, FRG

Bioassays provide inexpensive and easy-to-handle methods for the determination of phytotoxic compounds, mainly herbicides, in soil and water. Their advantages are the possibility of detecting only phytotoxic isomers, or to quantify residues and to determine possible recropping risks within the same bioassay procedure. It is disadvantageous on the other hand, that mixtures of phytotoxic compounds cannot be separated and that samples with an unknown history usually cannot be analysed properly.

Nevertheless, bioassays can be used for the quantitative determination of residues and give results comparable to physical-chemical methods. They are often sufficiently sensitive and especially suitable for screening procedures for new compounds where no physical-chemical methods have been developed yet. Many examples for different methods from the literature are given, and a bioassay procedure to quantify herbicide residues in water samples is described in detail.

Chemistry of Plant Protection, Vol. 11

1 Introduction

Bioassays in general are defined as experiments for estimating the potency of a chemical by analysis of the reaction that follows its application to living organisms [1]. Within the scope of this book a quantitative bioassay can be defined as an assay, where an unknown amount of a herbicide or other phytotoxic pesticide is measured by comparing the reaction that follows its application with the reaction caused by known standard rates of the same compound.

Bioassays can be classified as indirect methods for determining herbicide concentrations, which leads to a few limitations. Only phytotoxic compounds can be detected, therefore they are mainly applicable for herbicides. The interaction of herbicide and plant inherits the advantage of measuring the phytotoxic amount of herbicide, e.g. in the presence of non-phytotoxic isomers [2]. However, this can be detrimental, if the amount of the compound is overestimated due to phytotoxic metabolites, or if a mixture of chemicals is to be analysed. Most bioassay methods require a time of at least some days to show an adequate response, so that dissipation of the herbicide during the test period and thus a steady change of the concentration has to be taken into account. The range of sensitivity is often limited, since a close relationship between dose and response is shown only within a comparatively small range of concentrations.

Because of these limitations, bioassays have been regarded as insensitive, time-consuming and inconsistent during the period of fast improvement of chemical methods. However, following the invention of many highly active acetolactate synthase (ALS) inhibiting herbicides since the early 1980s reliable bioassays have been developed and have played a major role in recent years. This has shown that bioassays can provide low-cost and sensitive analyses, especially for highly biologically active compounds, for screening purposes and during the development of herbicides, while physical-chemical methods of analysis are not yet available.

A great variety of methods (see Table 1) have been described, depending on different purposes. The majority were used for the determination of remaining phytotoxicity in soil (carry-over), resulting in numerous assays for the quantitative determination of herbicides in soil, as compiled e.g. by Hurle [3] or Clay [4]. However, many of these methods can be applied with slight alterations to determine phytotoxic agents in water samples. Also methods which have not yet been used quantitatively could be adapted to the quantitative determination of phytotoxic agents, if the general conditions described below are fulfilled, therefore not only papers about quantitative bioassays have been reviewed.

In the following, the basic requirements which apply to bioassays in general, are described, as well as a short view on statistical evaluation, before methods for different groups of herbicides are specified. One example for the application of a quantitative bioassay method is described in detail.

Table 1. Overview of bioassay methods and their applications for main groups of herbicides with different sites of action

Type of assay	Response parameter	Site of action	Chemical classes	Duration	Reference
Algae or photo-autotrophic cell culture	O_2-development, cell counts, cell density, fluorescence	Photosynthesis phosphorylation cell division	Triazines, ureas, triazinones, pyrimidines, nitriles, chloroacetanilides, carbamates, diphenyl ethers, phenols	5 min–48 h	6, 24–34
Heterotrophic cell culture	Conductivity of growth medium	Nucleic acid metabolism	Sulfonylureas, imidazolinones	8d	24
Other short term assays	Leaf disk buoyancy	Photosynthesis phosphorylation radical damage	Triazines, ureas, triazinones, pyrimidines, bipyridinium compounds	30 min–24 h	35–38
	Chlorophyll fluorescence	Photosynthesis phosphorylation radical damage,	Triazines, ureas, triazinones, pyrimidines, bipyridinium compounds	8–48 h	39
	Tetrazolium reduction	Phosphorylation, fatty acid synthesis	Alkanoic acid esters, diphenyl ethers, oxims	2 h	40
Growth test-shoot uptake (floating or submerged)	Visual assessments, weight or length of whole plant or plant parts, chlorophyll–content area covered (Lemna)	Photosynthesis phosphorylation radical damage	Triazines, ureas, triazinones pyrimidines, bipyridinium compounds	10 min–72 h	24, 42–45
Germination test	Plant counts, radicle length, shoot length	Growth regulation, cell division	Alkanoic acids, amides, carbamates, chloracetanilides, dinitroanilines, oxims, thiocarbamates	2 to 4 d	24, 46–48 51–53
Shoot or radicle elongation	Shoot length, radicle length	Growth regulation, cell division and elongation, nucleic acid metabolism	Alkanoic acids, amides, carbamates, chloracetanilides, dinitroanilines, oxims, thiocarbamates, imidazolinones, sulfonylureas	2–6 d	16, 49, 50
Growth test-root uptake	Visual assessments, weight or length of whole plant or plant parts, chlorophyll–content	Photosynthesis, phosphorylation, radical damage nucleic acid metabolism	Phenols, triazines, triazinones, ureas, pyrimidines, bipyridinium compounds, sulfonylureas, imidazolinones	8–21 d	18, 22, 54–56

2 General Conditions for the Use of Quantitative Bioassays

2.1 Test Species

A suitable test species should show a close relationship between the amount of herbicide applied, and the reaction of the plant, with the required sensitivity. A fast development of freshweight and a manifestation of treatment effects at an early growth stage shortens the duration of the assay and keeps dissipation of the test compound to a minimum. A low inherent variability between replications and consistence between repeated assays even under slightly varying conditions will provide more reliable results. Weeds, though very sensitive, are often too variable in their reactions due to the wide range of genotypes within the species; the use of crop plants, especially F1-hybrids, results in a markedly lower variability [5]. The careful selection of uniform pregerminated seeds or seedlings can also lower variability considerably. Easy handling of the species as well as of gathering of results, e.g. plant counts, determination of fresh weight or leaf length, is an additional criterion. A great number of higher plant species have been used for herbicide bioassays. Compilations are given e.g. by Hurle [3], Kohn [6], or Parker and Ogg [7].

2.2 Comparability of Standard Curves and Unknown Samples

Untreated controls and standards with known concentrations of the test compounds have to be included in every quantitative bioassay with unknown samples. Several concentrations ranging from no effect to total damage are necessary to provide sufficient information to establish a reliable standard curve. The number of concentrations and replications per treatment depends on the expected reaction and its variability. If the range of concentrations leading to the desired effects is well known and consistent, fewer different treatments have to be included than for an assay with a compound with unknown or variable effects. Variability of the results will also require a higher number of replications. Values between 90% and 10% of the untreated controls are the most valuable for the determination of the dose-response relationship.

Bioassay indicators (especially higher plants) have a very limited response range, making it necessary to use different test species or dilute the samples to achieve reactions in the unknown samples within the valid range of the standard curve [8]. In many cases, e.g. after experimental applications of herbicides, the expected concentrations in the samples can be estimated and thus suitable dilutions be chosen. However, if no information is available on the concentration, a wider range of dilutions or preliminary screening tests have to be carried out. Sometimes it can be appropriate to use two different species of differing sensitivity to achieve a wider range of concentrations for quantification and to avoid dilutions [9].

It is important to consider possible differences between the unknown samples and the standard curve, especially where field samples are involved. Organic material and nutrient content have a great influence on herbicide activity [6]. It may be necessary to equalize the nutrient status by supplying concentrated nutrients to each sample. Standards made up from solutions similar to the samples are preferable, e.g. leachates from untreated plots or soil extracts rather than pure nutrient solution for the evaluation of leaching samples. As for the standard curve, the material used for dilutions should be similar to the samples.

2.3 Growth Conditions

The growth conditions should be close to the optimum for the test species. A fast growth provides a short duration of the assay as well as a higher sensitivity due to a higher intake of herbicides, since most herbicides are taken up by the plant passively with the water stream. During winter additional lighting may be necessary in glasshouses, but the results will still be more varying throughout the year than in growth chambers. For some plant species, e.g. *Lemna*, the day length is important for a vigorous growth, especially for stock cultures. The pH of the nutrient solution may not only influence the growth but also the uptake of the pesticide by the plants and thus the response.

2.4 Clean-Up and Extraction Procedures

Clean-up procedures for soil and water samples are usually not necessary for bioassays. Since a high content of organic material in water samples may lower the sensitivity of a test due to sorption, filtration may be required sometimes.

If the sensitivity of the bioassay methods available is not sufficient, it is also possible to use any chemical-physical extraction method, e.g. liquid-liquid partition or solid-phase extractions, to concentrate solutions. The method is applied in the usual way, and the solvent sample is then evaporated carefully to dryness. The remaining chemical can then be redissolved in water or nutrient solution at the appropriate estimated concentration, but the water solubility of the chemical is the limiting factor.

2.5 Statistical Evaluation

The relationship between plant response and herbicide concentration can be described as an S-shaped curve, which is almost linear around the point of inflection. A typical dose-response curve for the relationship between herbicide concentration and the growth of a test species is shown in an idealised form in Fig. 1.

The statistical treatment of the data should provide a reliable conclusion from the standard curve to the concentration in the samples, thus the application of

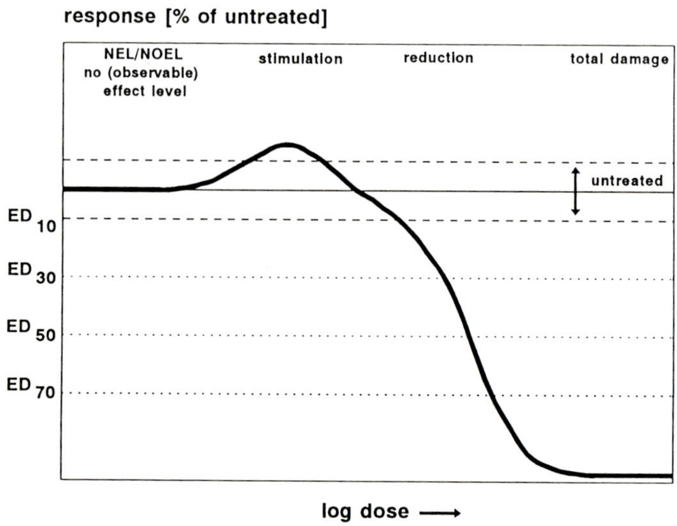

Fig. 1. Idealized dose-response relationship between an organism and a phytotoxic substance with range of variation (modified from Ref 8.)

regression procedures to the standards is the most appropriate one [e.g. 10,11]. Detailed descriptions of the mathematical procedures are given elsewhere [e.g. 12, 13], only a short view on possible methods in general is given here. Several approaches for the mathematical description of the relationship have been used, including transformations to achieve linearity [14–17] and nonlinear regressions [11,18–20]. The concentrations in the unknown samples are then quantified by comparing the data with the standard curve, the central part being the most reliable base. If the reaction is expressed as a percentage of untreated controls, the part of the standard curve between 40% and 60% [21] or between 30% and 70% of the controls [5] may be used. Below and above these values effect and concentration are not closely related. An example with standard curves for different herbicides is given in Fig. 2, where the activity to sunflower test plants is compared in water samples, according to the method developed by Günther et al. [22].

3 Methods for the Quantitative Determination of Phytotoxic Compounds

Bioassay methods can be classified by different criteria, e.g. duration (short- or long-term assays) or plant material involved [23]. The brief description of some methods given here is based on the experimental set up.

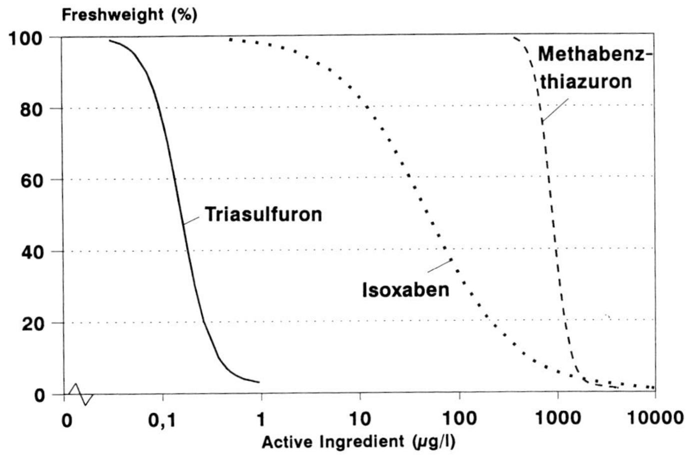

Fig. 2. Comparison of logistic dose-response curves for different herbicides with sunflower in water culture

Most bioassays used for the determination of phytotoxic compounds in water samples were in-vitro methods with algae, or aquatic plants were used. Growth tests with higher plants in solution or hydroponic culture were less frequent. Often algae or cell cultures are more sensitive, and they take less time than tests with higher plants, because the cells are equally exposed to the chemical, and cuticular penetration or translocation do not limit the amount of chemical entering the cells.

In general, the method to be used has to be related to the mode of action of the pesticide in question. Most of the assays described in literature were developed for photosynthesis inhibitors, especially many in-vitro methods. These herbicides, e.g. the triazines or the urea derivatives, cause many reactions in living plant cells which can be used as response parameters, such as a decrease in oxygen production or chlorophyll content but also less specific reactions like reduced growth. The latter is a reaction which is applicable for most compounds and can be measured e.g. as weight, size or density in cell cultures. Since photosynthesis inhibitors do not affect germination, assays involving the first growth stages before depletion of the seed reserves are not useful here. For compounds, on the other hand, which inhibit cell division, seeds, often pregerminated, are frequently used and show a quick reaction. The comparison of four different methods (cell and algae cultures, aquatic plants and germination) for herbicides from most chemical classes showed the applicability of the different methods for the main modes of herbicide action [24].

Table 1 gives an overview of the most important bioassay methods, and their applicability to several groups of herbicides with special respect to their mode of action.

In-vitro assays using algae like *Chlamydomonas* [25], *Chlorella* [26, 27, 28], *Dunaliella* [29, 30], *Scenedesmus* [24, 28] and *Selenastrum* [31], plant cell cultures [24, 32, 33], protoplasts [32] or isolated chloroplasts [34], are often shorter and considerably more sensitive than tests using higher plants. The results may be measured as cell counts [24, 25, 33], optical density [29], chlorophyll content [27, 30] or chlorophyll fluorescence [28, 33], as microscopic evaluations of cell shapes [29]. The growth can also be assessed by the conductivity of the growth medium which is inversely correlated with the increase in growth [24]. The O_2-evolution can be measured by electrodes [31, 33], as bioluminiscence of photobacteria [26] or in a Warburg apparatus. Since penetration and translocation have far less influence than within higher plants, they can be suitable for a wide range of herbicides [24, 25, 29]. Often they give results within a very short time, sometimes less than an hour [26, 28], usually within 24 or 48 h [24, 25, 27, 29, 31]. As a disadvantage, cell cultures of all kinds require the steady maintaining of stock cultures, highly controlled conditions and special equipment to achieve a low variability and repeatable results.

There are several other methods, all taking 48 h or less, which can hardly be grouped by another reason than by their short duration. In leaf disk buoyancy methods [35–38], the floating of punched leaf disks of various test plants in potassium buffer solution due to the oxygen developing during photosynthesis is assessed. They are suitable mainly for photosynthesis inhibitors and quaternary ammonium compounds. In a modification of this assay, the leakage of electrolytes from the tissue was measured [38]. The chlorophyll fluorescence in intact plants, which were transferred to herbicide solutions, changed within 48 h after transfer [39]. A quantitative relationship was found between the reduction of tetrazolium to formazan in *Sorghum* stem segments and the concentration of herbicides in the solutions where the stems had been incubated previously [40]. The direct measurement of in-vitro enzyme inhibition by herbicides is a relatively new development, but seems to be promising, because it needs only minute quantities of solutions and can be performed easily on a large scale [41].

Aquatic plants have been used in a number of laboratory assays, since they can be easily exposed to test solutions. Duckweeds (*Lemna* spp.) have frequently been used [24, 42, 43, 44], mainly for the determination of photosynthesis inhibitors, often taking advantage of the interference between paraquat-action and inhibition of photosynthesis [44]. Other aquatic plants used in similar assays include *Myriophyllum spicatum* [45] or *Elodea* spp. [6]. As parameters of effect, visual assessment [44], the digital recording of the area covered by the fronds [24], dry weight [42], changes in the conductivity of the ambient solution [43], or O_2-production [45] are applicable.

Assays with seeds can be used for herbicides which inhibit early growth stages, i.e. germination or shoot and radicle elongation. Examples are phenoxyacetic compounds, thiocarbamates, sulfonylureas or dinitroanilines. The seeds are exposed to the test compound usually in petri dishes on filter paper [46, 47, 48], quartz sand [49, 50, 51], agar [52], vermiculite [24] or in Erlenmeyer flasks in liquid shaking culture [53]. Frequently used test plants are garden cress [24,

47], maize [48], *Sorghum* [49, 50], cucumber [50, 53] and oats [16, 50]. Assays will last for 2 to 4 days and require a controlled temperature room. For volatile compounds, it is necessary to seal the petri dishes carefully [49]. Seeds may be pregerminated [16, 48, 50, 51], but this is not always necessary. Using pregerminated seeds it is possible to measure only root or shoot elongation [16, 49, 50], and variation will decrease, because uniformly developed seeds can be selected. Other response parameters which can be assessed are percentage of germinated seeds [52, 53] or the total length of the radicle [46–48], the hypocotyl [46] or the shoot [16, 52].

Growth tests with higher plants have not been used as frequently for the detection of herbicides in water samples as in-vitro methods. Most of them will last from 10 days to 3 weeks, depending on the test species, the herbicide and the growth conditions. The root is the most important way of uptake in these assays. Test plants may be sown, if the investigation of early growth stages is of interest; similar to germination tests. If photosynthesis inhibitors are under investigation, the test plant is preferably transplanted to shorten the assay by the phase until the seed reserves are depleted and the plant starts photosynthetic activity [16]. Numerous methods have been developed for the determination of herbicides in soil [3], but using water culture or a non-sorptive substrate such as quartz sand, some of them can be adapted to water samples, e.g. [18, 54, 55]. The plants can also be grown directly in the test solution [e.g. 22], or on agar plates [56]. Usually shoot fresh or dry weight [18, 22, 55, 56], root dry weight [54], shoot or leaf length [57] or chlorophyll content are measured. The tests are carried out in controlled environments such as glass houses or growth chambers.

Two new developments which use a specific biological reaction, biosensors and immunoassays.

4 Practical Use of a Bioassay Method for the Quantitative Determination of Sulfonylurea Herbicides

An example for the application of a bioassay method to the determination of sulfonylurea herbicides is briefly described in the following according to the method of Günther et al. [22]. As the first step, the most suitable test plant has to be determined. Important criteria are sensitivity and variability, as discussed in Sect. 2.1. Both can be influenced by the growth conditions, the test species and as well as the measured response parameters. Therefore it may be necessary to investigate more than one response parameter to find the most suitable one [22]. In Fig. 3, shoot freshweight data of three test plants (turnip, lentil, sunflower) are shown. Sunflower is not only more sensitive than the other two species, but also shows a lower variability and was therefore chosen as test plant for the determination of metsulfuron and triasulfuron in soil column leachates [22].

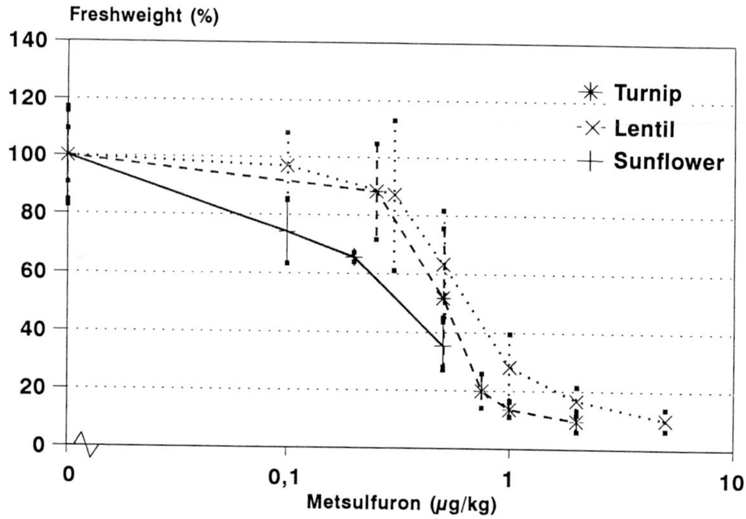

Fig. 3. Comparison of logistic dose-response curves obtained with different testplants for Metsulfuron-methyl in water culture

As shown previously, it is often necessary to dilute the water samples with an appropriate amount of a similar untreated control solution (see above). In this leaching experiment, there was not enough information about the behavior of the chemicals in question to estimate properly the concentration in the samples. Therefore, a first test was carried out with all the undiluted samples. Where the test plants were more than 70% smaller than the controls ($>ED_{70}$) the samples were bioassayed again with appropriate dilutions, together with a calibration curve. This was necessary for the samples from days 3 to 11. Dilutions of 1:5 and 1:10 were chosen, and the results are shown in Table 2.

Table 2. Quantification of metsulfuron-residues (µg/l) in soil column leachates. Limits of Quantification were 0.38 µg/l (ED_{30}) and 0.66 µg/l (ED_{70}) by using sunflower as test plant

Day	Dilution factor	Fresh weight (% of untreated plants)	Conc. in dilution from calib. curve	Conc. in sample
1	–	80.8	<0.38	<0.38
2	–	109.0	0	0
3	1:5	55.3	0.46	2,32
4	1:10	7.6	>0.66	>6.6
5	1:5	36.5	0,59	2,97
6	1:5	46.3	0,52	2,60
7	1:5	94.7	<0.38	<1.9
8	1:5	64.8	0.41	2.05
9	1:5	103.5	<0.38	<1.9
10	1:5	76.5	<0.38	<1.9
11	–	23.5	>0.66	>3.3

For all fresh weights between ED_{30} and ED_{70}, the corresponding concentration can be directly read from the computed dose-response-curve (see Fig. 3). For the sample from day 4, the dilution of 1:10 was not sufficient (fresh weight 7.6% of untreated plants), therefore it can only be estimated, that the concentration in the original sample was higher than 6.6 μg/l, ten times (the dilution factor) the ED_{70}-value. For samples from days 7, 9 and 10 on the other hand, the dilution 1:5 was too high, therefore it can only be estimated that the concentration was less than 1.9 μg/l, five times the ED_{30}-value. Since there were not enough samples left for another bioassay, a final quantification was not possible.

5 Comparison Between Bioassay and Physical-Chemical Methods

For many groups of herbicides, e.g. triazines or ureas, the detection limits of analysis by GC or HPLC exceed the phytotoxic limits by far [55, 57]. Methods are well standardized and comparatively easy, achieving a high number of sample workups per day. Herbicide mixtures are relatively easy to analyse, even several compounds within one step. However, these sophisticated methods require very expensive equipment which needs regular maintenance, a high technical knowledge, and expensive materials.

Compared to this, most bioassay procedures are slow (up to several weeks) and herbicide mixtures or samples of unknown origin are very difficult to handle. On the other hand, clean-up procedures of samples are usually not necessary, and the equipment for many assays is cheap and easy to maintain, requiring fewer skills from the staff. Another great advantage is the possibility of getting information on the residue level as well as on the potential phytotoxicity, because only the biologically active material is measured.

6 Outlook and Conclusion

Especially for the highly biologically active groups of sulfonylureas and imidazolinones, bioassays are still the standard method of quantitave determination [58]. Advantages of using a bioassay for estimating residues of these herbicides are the extreme sensitivity of a number of plant species. For these herbicides, it is very important that bioassays can provide the quantitative determination as well as the risk assessment for succeeding crops within one step. Bioassays for quantitative analysis are also an option, where expensive laboratory equipment and highly skilled personnel are not available but suitable growth conditions for a bioassay can be provided.

With environmental concerns about pesticide use gaining more and more importance, bioassays will still be necessary in the future because additional information is needed on possible impacts on the environment in addition to the amount of the total residues. Unlike bioassays, instrumental methods give no indications of possible hazards be they for crops or non-target organisms.

7 References

1. Streibig JC (1992) Quantitative assessment of herbicide phytotoxicity with dilution assay. Thesis, Royal Veterinary and Agricultural University, Copenhagen
2. Chaloner PA (1989) Proc. Brighton Crop Prot. Conf. Weeds Vol. 2: 697
3. Hurle K (1977) In: McFarlane NR (ed) Crop protection agents – Their biological evaluation, Academic Press, p 285
4. Clay DV (1993) In: Streibig JC, Kudsk P (eds): Herbicide bioassays. CRC Press, Boca Raton, FL, p 153
5. Pestemer W (1976) Weed Res. 16: 357
6. Kohn GK (1980) Res. Rev. 76: 100
7. Parker R, Ogg AG Jr (1990) Extension-Bulletin Cooperative Extension, College of Agriculture and Home Economics, Washington State University, EB 1417
8. Beyer EM, Duffy MJ, Hay JV, Schlueter DD (1987) In: Kearney PC, Kaufman DD (eds) Herbicides: chemistry, degradation, and mode of action, Marcel Dekker, New York, p 117
9. Iwanzik W, Amrein J, Koeller R (1988) Z. Pflanzenkr. Pflanzenschutz, Sonderheft 11: 301
10. Cousens R, Marshall C (1987) Ann. Appl. Biol. 111: 469
11. Streibig JC (1988) Weed Res. 28: 479
12. Draper N, Smith H (1981) Applied regression analysis, 2nd edn. John Wiley, New York
13. Streibig JC, Rudemo M, Jensen JE (1993) in: Streibig JC, Kudsk P (eds): Herbicide Bioassays, CRC Press, Boca Raton, FL, p 29
14. Allott DJ, O'Neill JA (1970) Rec. Agrig. Research Minist. Agric. North. Ire. 18: 21
15. Beckett PHT, Davis RD (1977) New Phytologist 798: 95
16. Nyffeler A, Gerber H-R, Hurle K, Pestemer W, Schmidt RR (1982) Weed Res. 22: 213
17. Pestemer W (1983) Berichte Fachgebiet Herbologie Universität Hohenheim 24: 85
18. Günther P, Rahman A, Pestemer W (1989) Weed Res. 29: 141
19. Streibig JC (1980) Acta Agric. Scand. 30: 59
20. Brain P, Cousens R (1989) Weed Res. 29: 93
21. Gerber HR, Ebner C, Frei P, Bieringer H (1975) Z. Pflanzenkr. Pflanzenschutz, Sonderheft 7: 169
22. Günther P, Pestemer W, Rahman A, Nordmeyer H (1993) Weed Res. 33: 177
23. Pestemer W (1979) Z. Naturforsch. 34C: 964
24. Grossmann K, Berghaus R, Retzlaff G (1992) Pest. Sci. 35: 283
25. Hess FD (1979) Abs. 178th Nat. Meeting Am. Chem. Soc.: 73
26. Tchan YT, Roseby JE, Funnell GR (1975) Soil Biol. Biochem. 7: 39
27. Lefebvre-Drouet E, Calvet R (1983) Berichte Fachgebiet Herbologie Universität Hohenheim 24: 63
28. Schmidt C (1983) Ecotoxic. Environ. Safety 7: 176
29. Felix HR, Chollet R, Harr J (1988) Ann. Appl. Biol. 113: 55

30. Yarden O, Freund M, Rubin B (1993) Fresenius Environmental Bull.2: 31
31. Bennet PH, de Beer PR (1984) Pest. Sci. 15: 425
32. Pallett KE, Rees RT, Fitzsimons PJ, Cobb AH (1986) Asp. Appl. Biol. 11: 139
33. Thiemann J, Nieswandt A, Barz W (1990) Plant Cell Reports 8: 399
34. Hirschywald B, Zimmermann GM, Normann S, Haberer K (1990) Z. Wasser-Abwasser-Forschung 23: 184
35. Truelove B, Davis DE, Jones LR (1974) Weed Sci. 22: 15
36. Saltzman S, Heuer B (1985) Pesticide Sci. 16: 457
37. Petzold U, Wichmann R, Dobe H (1988) Arch. Phytopathologie Pflanzenschutz 24: 445
38. Yanase D, Andoh A, Chiba M, Yoshida S (1993) Z. Naturforschung 48C: 397
39. Shaw DR, Peeper TF, Nofziger DL (1986) Crop Sci. 26: 756
40. Rehab IF, Bendixen LE (1989) Special-Circular, Ohio Agricultural Research and Development-Center No. 125: 42
41. Wiegand-Rosinus M, Obst U, Haberer K, Wild A (1992) Env. Toxicology, and Water Quality 7: 313
42. Damanakis M (1970) Weed Res. 10: 77
43. O'Brien MC, Prendeville GN (1978) Weed Res. 18: 301
44. Parker C (1965) Weed Res. 5: 181
45. Selim SA, O'Neal SW, Ross MA, Lembi CA (1985) Proc. North Central Weed Control Conf. 40: 40
46. Reid CPP, Hurtt W (1969) Weed Res. 9: 136
47. Neururer H (1975) Z. Pflanzenkr. Pflanzenschutz 82: 316
48. Morishita D, Thill DC, Flom DJ, Campbell TC, Lee GA (1985) Weed Sci. 33: 420
49. Parker C (1964) Proc. Brit. Weed Contr. Conf. 7: 899
50. Kratky BA, Warren GF (1971) Weed Res. 11: 262
51. Blacklow WM, Pheloung PC (1992) Australian J. Agric. Res. 43: 1157
52. Dornbos DL Jr., Spencer GF (1990) J. Chemical Ecology 16: 339
53. Kuboi T, Fujii K (1984) Soil Sci. Plant Nutr. 30: 209
54. Damanakis M, Drennan DSH, Fryer JD, Holly K (1970) Weed Res. 10: 278
55. Stalder L, Pestemer W (1980) Weed Res. 20: 341
56. Pestemer W, Nolting HG, Lundehn JR (1979) Nachrichtenbl. Deut. Pflanzenschutzd. 31: 166
57. Horowitz M (1976) Weed Res. 16: 209
58. Iwanzik W (1989) Proc. Brighton Crop Prot. Conf. Weeds Vol. 2: 697

Assurance of Data Quality

Hans Egli

Crop Protection Division, Ciba-Geigy Ltd., CH-4002 Basel, Switzerland

Published data and analytical experience reveal an inter-laboratory "basic variability" inherent in pesticide trace analysis in water amounting to ±50% (95% confidence level). This favourable picture is unfortunately often obscured by erratic, false results deviating far more from the true values. Those results must be attributed to gross errors resulting from inadequate analytical practice. Failure to confirm results where indicated is one of the most frequent gross errors. Results of acceptable quality need not only be correct within the "basic variability", they must be supplemented with an estimate of the accuracy, and a comprehensive documentation covering the whole data generation process. The use of suitable analytical methods in an appropriate manner is a prerequisite. Although principles of "Good Laboratory Practice" do not guarantee by themselves correct results, they contribute to data quality if applied in a scientifically sound way. Relevant items of GLP are: Comprehensive study plans designing and limiting the whole study, including site selection and sampling; "Standard Operating Procedures" addressing all generally critical points in the analytical method, defining that and how certain elements such as calibration, calculations, recovery experiments etc. are carried out; and a documentation plan assuring that the data generation process remains transparent.

Chemistry of Plant Protection, Vol. 11

1 Introduction

Any analytical result is expected to meet the true value within certain limits. To know these limits and to keep them acceptably narrow, is what is understood by quality of data. All measures to control the factors having impact on quality and to evaluate quality are included in quality assurance.

This chapter deals with the present state of quality in the area of pesticide trace analysis in water and the possibilities of improving it. Its purpose is to provide some ideas on the variability inherent in trace analysis and to give a logically and scientifically sound, but simple and practical concept, to obtain high quality results. Emphasis is laid on GLC and HPLC, since these methods are still the techniques in most wide-spread use in this domain.

Quality is not only a matter of concepts and techniques. Above all it has to do with the analyst who must not only have a thorough understanding of what she or he is doing, but must also have the will to do a good rather than a quick job. Additionally it requires a certain generosity, because high quality results do not mean ultimate security about the last digit. Analysts aiming at this goal do not work thoroughly, they are simply not able to recognize what is important. Quality in analytical chemistry is a sensitive balance between generosity and pedantry.

2 Performance of Trace Analytical Methods

There are many sources of information contributing to our knowledge about the performance of trace analytical methods with respect to accuracy, i.e., the deviation of analytical results from the true values: Collaborative studies, validation studies, tests of commercial laboratories, and – last but not least – many years of analytical experience by each analytical laboratory active in this field. The combination of all these pieces of information gives a rather consistant – although rough – estimate of an inherent "basic variability" in pesticide trace analysis in water. As will be shown, this "basic variability" is of course exceeded if unsuitable methods are used or if gross errors are made.

2.1 Collaborative Studies

In the last couple of years, several collaborative studies of pesticide trace analysis in water have been published by various authors. Table 1 shows the results of 7 studies including various pesticides, methods, and numbers of participating laboratories and covering a wide concentration range. The results are not fully comparable because the statistical treatment of the data was not the same in all

Table 1. Results of collaborative studies of pesticides in water

Compounds	Number of laboratories	Results	Range [$\mu g/l$]	Mean recovery	RSD	Outliers	Ref.
2[a]	6	592	0.5–500	101%	15%	37 (included)	[1]
10[b]	8	960	5–50	98%	7%	138 (excluded)	[2]
45[c]	10	5400	2–50	94%	19%	869 (excluded)	[3]
2[d]	5	100	5–5000	83%	12%	none	[4]
29[e]	11	3828	0.25–25	94%	18%	490 (excluded)	[5]
13[f]	5	130	3–30	81%	35%	many (incl.)	[6]
10[g]	13	220	0.04–0.8	100%	23%	excluded	[7]

RSD Relative standard deviation (inter-laboratory)
[a] Glyphosate and aminomethylphosphonic acid, HPLC determination (Fluorescence detector)
[b] N-Methylcarbamoyloximes and *N*-methylcarbamates (Fluorescence detector)
[c] Nitrogen- and phosphorus-containing pesticides, Capillary-GLC (P/N-detector)
[d] Glyphosate and aminomethylphosphonic acid, HPLC determination (Fluorescence detector)
[e] Organochlorine pesticides, Capillary-GLC (Electron capture detector)
[f] Organochlorine pesticides, Capillary-GLC (Mass spetrometric detector)
[g] Triazines, ureas, phenoxypropionic acids and other herbicides, Capillary-GLC (P/N-detector), HPLC (UV-detector), TLC

studies, and the outliers were included or excluded. Nevertheless the studies give a fairly consistent picture of the accuracy.

Most of the studies resulted in high recoveries and rather good precision. In the one exception [6], the authors identified poor GLC performance and several gross errors as the main sources of the poor data quality. There is no remarkable concentration dependence of the precision over a wide concentration range in these studies.

2.2 Other Studies Related to Accuracy

A comprehensive method validation test was organized and reported by US–EPA and Batelle [8]: 104 compounds were determined using 6 methods in fortified samples, a total of 3221 analyses resulted in an average recovery of 95% with an mean standard deviation of 10%.

In the Phase 1 Report of the "National Survey of Pesticides in Drinking Water Wells" [9] the mean recoveries for 106 compounds found by contract laboratories were 95% and the mean standard deviation was 12%, in the EPA laboratories the mean recoveries found were 91%, the mean standard deviation 11%. Another 5 compounds have not been included into this calculation because of obvious failure of the method (recoveries below 50%).

In the Basel residue laboratories of Ciba water samples have routinely been analyzed for many years. 873 recovery values out of the analytical reports from

1977 to 1991 revealed a mean recovery of 95% with a standard deviation of 14%. The recoveries were roughly normally distributed, thus showing that the factors influencing the results are of essentially random type. The compilation includes results for 30 pesticides and metabolites, including triazines, triazole fungicides, chloroacetanilides, organophosphates, phenylureas, sulfonylureas, and others, obtained by numerous analysts and various methods (GLC, HPLC).

2.3 An Estimate of the Accuracy

In chemical analysis, the overall-accuracy is usually split into random errors or precision, and systematic errors or bias. Precision refers to repeatability and reproducibility, in intra-laboratory analyses errors of this type are caused by apparently random factors which are not under the control of the analyst, but which average out after a sufficient number of repetitions. Bias in intra-laboratory analyses includes factors reproducibly influencing the result in distinct directions, such as false reference concentrations, inadequate calculation procedures, interfering compounds in a reagent. If analyses are repeated in other laboratories, the intra-laboratory random factors remain principally the same, although some of them may be under better control in other laboratories and therefore change their importance, whereas those intra-laboratory biases which are rather laboratory-specific than method-specific, become random errors at the inter-laboratory level. It is thus obvious that it is difficult to make a clear distinction between precision and bias in trace analysis on an inter-laboratory level. Such a distinction would require large numbers of repetitions, and the conclusions from these would be difficult to generalize and thus be of very limited value for analytical practice. For a single result the important figure is the deviation from the true value, irrespective whether the error is systematic or random. Therefore accuracy is understood to include both precision and bias. A generally valid estimate of the accuracy cannot be given, instead a rough "basic variability" is derived.

The random error found in the studies and compilations cited lies between 7% and 23% if outliers are not considered. The average recovery values are usually between 90% and 100%, corresponding to a relative average systematic error of between 0 and 10%. If a normal distribution is assumed, bracket values can be estimated which include a certain percentage of all measured replicates of a given sample by various laboratories. Bracket values including 95% of all results (b.v.) are described by Eq. (1).

$$\text{b.v.} = (\text{mean value}) \pm 2(\text{RSD}) \quad (1)$$

As can be seen from the bracket values derived from the studies described in paragraphs 2.1 and 2.3 (Table 2), all but one are well within 50% and 150%. From these rather consistent data it is therefore concluded that there is a "basic variability" of pesticide trace analysis in water of ±50% expressed as bracket values.

Table 2. Bracket values

Mean recovery	RSD	Bracket values mean ± 2 (RSD)
101	15	71–131
98	7	84–112
94	19	56–132
83	12	59–107
94	18	58–130
81	35	gross errors
100	23	54–146
95	10	75–115
91	11	69–113
95	12	71–119
95	14	67–123

The "basic variability" includes both intra- and inter-laboratory deviations, random and systematic errors. The intra-laboratory variability may in some cases be smaller, e.g. if a laboratory is well-trained on a specific method. Nonetheless, with the results taken as a whole, ±50% is a realistic estimate of the "basic variability". It includes a certain allowance for additional errors in the analysis of real test samples as compared to collaborative studies and recovery experiments actually representing ideal situations. It should be noted, however, that this number is only valid in the concentration range between 0.05 μg/1 and 5000 μg/1. A recent collaborative study of chlorinated dibenzo-p-dioxins and dibenzofurans revealed that in the sub-ng/l range the accuracy is worse [10]. Mainly due to laboratory-related systematic deviations, analyses at that level are at or beyond the limit of reasonable reliability.

The proposed "basic variability" is much better than that to be expected according to Horwitz [11]. Based on the Horwitz curve (dependance of the precision on the concentration) the accuracy of pesticide trace analysis in water in the sub-μg/l- range is supposed to be at least ±60% standard deviation which would correspond to a 95% confidence range (comparable with the "basic variability") of about ±120%. With this high variability all results would include the zero-value as possible true value. Single measurements would thus be below the actual limit of determination which must be significantly different from the zero-value. The "basic variability" differing from the Horwitz value can be explained by: (1) precision is not merely a question of concentration, but also of the amount of coextractives, these are much lower in water than in other substrates, (2) because the limits of determination depend on the precision, the lowering of their limits over the last decade was only possible after improvement in the precision. The Horwitz curve which relies on data of the 1960s and 1970s is thus not directly applicable.

2.4 And The Reality?

The "basic accuracy" as elucidated in the last paragraphs suggests that traces of pesticides in water can be determined with reasonable and acceptable reliability. No more than 5% of the values would be allowed to be outside ±50%. Nevertheless, several ring tests carried out among commercial laboratories revealed another reality.

In 1986 a multilaboratory study among 6 independent laboratories was organized by Health and Welfare Canada [12]. Spiked water samples were analyzed for chlorinated aromatics, organochlorines, polycyclic aromatics, and organophosphates by a GCMS multicomponent method. The spiking levels were in the range of 2 to 15 μg/l. The deviations of the results from the nominal values were laboratory-dependent, on the average, 21% of the values were outside ±50% of the nominal values.

The US Department of Agriculture submitted streamwater samples fortified with picloram and 2,4–D to 10 commercial laboratories [13]. Spiking was at levels of 10 and 50 μg/l. As many as 60% of the values were outside ±50% of the nominal ones.

Another test was sponsored by Ciba [14,15]. Five water samples were submitted for determination of atrazine to 11 commercial laboratories. Four of the samples did not contain any atrazine, one was spiked with 0.1 μg/l atrazine. Totally 47% of all values were either outside ±50% of the nominal values or were falsely positive.

These results are shocking for analysts devoted to good science. Is the "basic variability" so much less favourable in reality? A closer look at the results of these ring tests clearly shows them to be rather erratic than randomly distributed. The "basic variability" however, includes only random errors and minor systematic ones. The results of these test analyses are thus simply false because of gross errors, i.e. avoidable errors which result from inadequate analytical practice including failure to confirm the results.

In reality, there are obviously two ways of practicing pesticide trace analysis in water. Following the better one, well-trained and science-oriented analysts correctly use valid methods, critically review and confirm their results before publication; as a consequence the results are accurate within the "basic variability". The "quick-and-dirty" practice unfortunately still continues to exist: Inappropriate methods are applied or basically valid methods are misused, first results are released without self-criticism and without confirmation; as a consequence erratic results are produced.

2.5 Gross Errors

Examples of frequent gross errors are losses/contamination during storage or cleanup of the samples, operational mistakes during analysis, use of unsuitable reagents, inappropriate chromatographic conditions, instable chromatographic systems, misinterpretation of chromatograms, calculation errors. At low concentration levels gross errors occur more often, because many of the frequent ones are not concentration-dependent but of absolute type, thus influencing low-level results more strongly. Likewise, false positive and false negative values are more likely to be found at low concentrations.

Gross errors are usually laboratory-related, not method-related, although some methods might be more prone to gross errors than others. Gross errors result from inadequate use of the method or from lack of experience. As mentioned above failure to perform confirmation analyses where indicated is nothing less than a gross error, perhaps the most frequent one. Gross errors do not happen, they are committed. They can be avoided by adequately trained people, by strict observation of precautionary measures, and by critical examination of the results.

Such errors are not statistically controlled, and thus cannot be reasonably included into statistical discussion of data quality. They superimpose on the "basic variability" yielding erratic, false results. Often they are laboratory-related systematic errors resulting in one laboratory's constantly gross over- or underestimation of the concentration. In cases where gross errors occur at random, the analyst is obviously not in control of the analytical method.

In the collaborative studies discussed in Sect. 2.1, up to 16% "outliers" were discarded. This is well in-line with former findings [16]. Outliers are nothing other than false results caused by gross errors. Or, grossly false results are euphemistically called outliers. Of course, outliers are a part of the real world, nobody is perfect. But the analyst can and must aim at avoiding them and has to give evidence that the results are likely to be free of them.

3 How Analytical Results of Acceptable Quality are Obtained

It is generally recognized that the intra-laboratory precision accounts for one-half to two-thirds of the inter-laboratory variability [16], because the analyses within one laboratory do not reveal laboratory-related systematic errors. The "basic variability" was shown to be ±50%, the corresponding standard deviation is ±25%, if the "basic variability" is taken as a 95%-confidence limit of a normal distribution. The intra-laboratory contribution then is about ±15%, being in good agreement with the percentages reported in Sect. 2.2. The result of a pesticide trace analysis in water is thus acceptable if its intra-laboratory precision does not exceed ±15% standard deviation. Together with a laboratory-related bias of about ±10%, the overall-accuracy amounts to ±25% standard deviation. In these figures, no allowance is included for gross errors.

Well-designed and valid methods are a pre-requisite for producing results within the "basic variability". A method is a written procedure describing how the analytes are determined. It must provide sufficient details about all steps as to allow a well trained analyst to perform the analyses. It must list the equipment needed or suitable substitutes. A method must be rugged and all critical steps have to be identified, it must allow qualified analysts to generate reproducible results. There must be a rationale behind each step of a method which should be understood by the analyst. Suitable control measures to improve – and to check – accuracy and to avoid gross errors have to be described. A method must be valid, i.e. there must be evidence that the method works as it is claimed to do.

In trace analysis there is no such thing like an "approved" method automatically yielding correct results. Such methods are only available for substrates of more or less constant composition. At the trace level, water samples of different origin have no constant composition, the amount and nature of constituents having a potential influence on the results vary. The analyst must critically review the results, even if using validated methods.

Not only should correct results be reported, but at least a rough estimate of the accuracy of the result has to be given, mainly to prevent over-interpretation of results or differences in results. As will be discussed later main sources for such an estimate are calibration and recovery values. If indicated confirmation analyses have to be performed. Records of all pertinent details of the analytical process must allow the results to be reconstructed. This documentation has to include all steps from sampling to the calculation of the results.

During the last some fifteen years, the concept of "Good Laboratory Practice" (GLP) has been set up in order to provide a formal basis for reliable results. Many of the before-mentioned requirements for acceptable data can be enforced by establishing an appropriate GLP-system. There are two aspects, however, which cannot be over-emphasized:

- GLP principles even if adapted to the special problems of pesticide trace analysis in water in a scientifically sound manner do not guarantee that the results are correct.
- GLP guidelines can be formally followed without having great impact on data quality. An analytical laboratory may well be in-line with GLP-guidelines without respecting traditional good science. Many of the laboratories participating in the ring tests described in Sect. 2.4 were working "under GLP"; the results clearly illustrate that this kind of GLP did not essentially contribute to good data quality.

Nonetheless, GLP-principles are a helpful tool to improve data quality if not used to replace rather than to reflect science. Today GLP has to be observed when generating data intended for registration of pesticides and many other regulatory purposes. Although monitoring studies are, as yet, excluded from GLP, its principles should apply as well if results of such studies are used as a basis for regulatory decision-making.

4 Merits and Limits of Good Laboratory Practice

The basis of GLP-regulations are guidelines: Derived from OECD-guidelines [17] there are many national ones. All include similar GLP-principles aimed at transparent generation of high quality data, proper documentation and archiving of raw data and other pertinent information, as well as at quality assurance by personnel not involved in the study. The whole system comprises many rather formal elements such as: Organization of test facilities, qualification and education of the personnel, availability of adequate equipment, handling of reference compounds and reagents, study plans, standard operating procedures, quality assurance unit, documentation, archives. Most of these elements are not directly applicable to the analytical practice; they have to be interpreted and adopted to the specific case.

The obligation to interpret the guideline elements is both an opportunity and a challenge. If interpreted by science-minded people, guidelines, establish and enforce good science in a laboratory, analysts are made aware of critical points in their daily work and are encouraged to think critically about what they are doing. If they are interpreted by formalists focussing to the letter rather than to the spirit, GLP degenerates into a purely formal system lacking any scientific basis. Such a formal GLP-system makes sure that raw data are recorded, that the results can be re-examined even after many years, that deliberate fraud is impeded. But beyond this it does not add much to the data quality. "Good laboratory practice" does not automatically mean good analytical practice [18].

It is outside the scope of this article to discuss in depth all the elements of GLP with respect to pesticide trace analysis in water. The discusssion is limited to study plans, standard operating procedures, and documentation. Other elements are of less or less direct importance for the data quality. Furthermore, no formal requirements will be detailed, only parameters and conditions having an essential impact on the result will be discussed.

4.1 Study Plans

The results are never interesting just per se, they are rather a part of a more comprehensive research study used for regulatory decisions, to answer questions about the environmental behaviour of a pesticide, to verify or falsify models, to monitor real situations or for other purposes. Ideally, the analyst not only determines concentrations in water samples which have arrived in the laboratory, but also participates in the overall-study and keeps its goal in mind while designing and performing the analyses, reviewing the results, and writing the reports.

It is a fundamental requirement of GLP that study plans be available before the start of the study. Such plans must include the objectives of the studies, the questions to be answered, the intended use of the results, parameters to be measured or otherwise collected, procedúres and evaluation methods. Although the system looks rigid it is actually not. If new aspects arise during the course of the study, deviations from the original plan are possible. However, deviations have to be treated as such: They have to be amended in writing to the study plan, the reasons for the deviation must be given.

In pesticide trace analysis in water the study plan has to mention the intended analytical methods. It must define which analytical information is necessary, e.g. analytes to be determined, limit of quantitation to be envisaged. It must describe the quality control measures. Of course, the study plan can make reference to written documents such as methods or standard operating procedures (which are discussed in the next paragraph).

The study plan should also cover the study design, sampling sites, and sampling procedures which have to be adapted to the objective of the study. It is important that every study is well designed, i.e. that the questions to be clarified by the study are asked *before* the start of the study and are mentioned in the study plan. The purpose of the study must be clearly stated and the conclusions being drawn must remain within this purpose. It is not scientifically appropriate to generalize findings of special studies; the conclusions drawn would be incorrect, even if based on otherwise correct and confirmed results.

The protocol has also to mention all pertinent information needed to interpret the results (even if not avilable yet at the stage of planning), such as hydrological data, topographical information, soil type, history of pesticide use, climate and weather.

Obviosuly, the design of a study is a typical multidisciplinary task. Depending on the scope of the study, besides analysts there should also be agronomists, hydrologists, meteorologists, soil scientists or others involved.

4.2 Standard Operating Procedures (SOP)

Repetitive and sensitive steps should be described in writing as so-called SOPs which are a kind of an agreement or directive in an analytical facility describing how a certain step of the analytical procedure has to be carried out. All generally critical points of the analytical practice should therefore be covered by a pertinent SOP. Because the laboratory management has to sign them, the SOPs obtain an official and obligatory character. Well designed SOPs are of utmost importance for the analytical process and data quality.

GLP does not tell the analyst which details of the procedures have to be covered, neither how they are structured. When creating SOPs it is more important than everywhere else in GLP that science is added to formalism. Here, factors having the greatest impact on the accuracy of the analytical results can be addressed; by SOPs, quality control measures are enacted; in SOPs good science is formalized.

In the following paragraphs all the items which can strongly influence the accuracy or even lead to gross errors are discussed. Many of these steps may be described in the analytical method, otherwide they should be addressed in SOPs. Incorporating the overall goal of accurate results, SOPs should give details of properly established and approved laboratory practices which guaranteee that each step of the procedures (from sampling to reporting) is under control. Usually there are many different ways of doing things correctly, the hints given below are mainly based on practical experience, some on published papers.

4.2.1 Sampling and Sample Storage

Sampling is a sensitive and still a problematic area. Because it depends so much on the water body and the individual study objective, no general sampling procedures can be given. Valuable hints for sampling of ground and surface water can be found in [19] and [20]. If repetitive studies (i.e. of the same type) are carried out, pertinent SOPs should describe how to take samples, assuring that sampling is always performed in the same way.

Glass bottles are best suited for water samples. On glass even more so on plastic surfaces (including PTFE) some pesticides mainly the unpolar ones can be adsorbed [21]. The suitability of the vessel material to be used for the samples must therefore be checked with all analytes in question. Besides causing losses through adsorption unsuitable plastic containers can release organic constituents interfering with one or the other analyte.

Although samples should principally not be stored for an extended period of time, storage may be necessary for reasons of temporary work overload or common analysis after the last sampling of a study. Hydrolytically stable compounds are usually stable on storage [22]. It is still questionable whether samples should be frozen for storage. Storage temperatures above 0 °C may cause microbial breakdown whereas freezing of the samples represents a certain risk of irreversible coflocculation or co-precipitation of pesticide traces with natural water constituents [23]. If the storage stability is not known it has to be checked under the chosen conditions.

The use of preservatives is recommended only if proven to be necessary and to have no negative impact on the analytes. Mercury salts used as preservatives can lead to analytical interferences [24].

4.2.2 Extraction and Cleanup

The sample preparation procedures are part of the analytical methodology. Their appropriate performance has to be regularly checked by recovery experiments (cf. Sect. 4.2.7.).

Usually one or more steps involve the measurement of volumes (ailquoting, final volumes). Those volumetric steps having a direct impact on the result must be carrried out with calibrated equipment. A precision of ±2% is proposed as sufficiently valid. This point needs careful consideration in SOPs. Special care has to be taken when handling volumes of below 1 ml. If pipettes, volumetric flasks, graduated cylinders, and/or syringes of volume-certified type are used the precision is guaranteed over a certain range. However, it has to be borne in mind that the calibration of the devices might not be sufficiently valid over the whole graduated volume range: A volume of 1 ml measured in a graduated 50 ml-pipette is likely to lack sufficient precision.

If the volumetric devices are either not of certified type or used beyond the certified range, they have to be calibrated. The calibration is best done by weighing a measured aliquot of water. The device and the procedure is acceptable if the average deviation from the nominal values of 20 replicate measurements is better than ±2%.

4.2.3 Final Determination

The final determination is the step in trace analysis most strongly influencing the overall-accuracy. There are many potential sources of errors which need to be observed and controlled. As already mentioned quantitation techniques other than gas or liquid chromatography are not considered here. The relevant components of the chromatographic system to be separately discussed are the injection system, the separation system, and the detection system (including the registration of the signal). The calibration as a performance test of the final determination system is dealt with in Sect. 4.2.4.

Injection system. Injection volumes need not necessarily be accurate, the reproducibility of the volumes, however, must be within ±2%. Reproducible deviations from the nominal values are averaged out if the same injection procedure is used for calibration. Reproducibility is no problem with loop injectors in HPLC, in GLC, however, manual injection essentially contributes to the overall-error, the use of mechanic injection devices (autosamplers) is therefore recommended. More or less pronounced carry-over effects often occur in chromatography. They are usually caused by the injection system. Tests have to be carried out regularly to show the absence of relevant carry-over effects. This is best done by injection of a rather concentrated standard solution followed by a check solution. Minor carry-over effects of up to 0.5% can hardly be avoided. This means that rinse injections have to be made whenever such carry-over would be relevant, e.g. after injection of highly concentrated solutions.

Chromatographic system. The separation of the analytes from each other and from interferences depends on the quality of the stationary phase, the stability of the mobile phase flux, and on the equilibrium between the two phases. This latter factor is of special importance if a gradient elution technique is being used. Instable chromatographic systems manifest themselves through varying retention times. It is therefore necessary to watch the retention times of standard injections and recovery injections very carefully. Acceptance criteria should be stated based on experience; in stable chromatographic systems, retention times do not vary by more than a few seconds. The retention window as formed by the measured retention times is subsequently used for identification of the peaks (cf. Sect. 4.2.5.), too wide a window therefore results in an enhanced risk of false positives.

Detector. Varying responses reveal unconditioned or instable detectors. Among the factors having impact on the detector stability, the sample matrix often plays an important role. It is therefore appropriate to condition the detector with a repeated injection of a recovery sample and to start the analytical run only if the analyte peaks of all injections do not vary by more than 5%. It is also important to prove the detector stability throughout the analytical run, this is best accomplished by evaluating the variations of the standard injections.

4.2.4 Calibration

Basically the whole analytical procedure including all extraction, cleanup and concentration steps needs calibration. It has become general practice to simplify this whole process by reducing calibration to the calibration of the chromatographic measuring system and to perform the calibration of the whole process by recording a "recovery function" or – more often – by running recovery samples.

The calibration of the chromatographic system by injection of standard solutions, i.e. of solutions of known content of analytes, is used to transfer the response of a sample injection into the result by a mathematical algorithm. There are many different ways of calibration in practical use: Internal reference, single external standards, external standard curves (graphic plot), external standard curves (various regression procedures). The calibration procedure should not only yield the result in a transparent way, but the analyst should also select it based on scientific arguments aiming at obtaining as much information about the analytical system as possible.

What purpose should calibration be designed for? (1) to obtain the basis for a statistically sound algorithm to calculate correct results, (2) to reliably estimate the random variance of the chromatographic system, (3) to define the valid working range including the lower limit of quantitation (cf. 4.2.6), (4) to make appropriate acceptance criteria available, and (5) to verify a sufficient stability of the chromatographic system. In order to cover all aspects it is recommended to inject a sufficient number of external standard solutions of suitable concentrations and to calculate standard functions for each analyte using an appropriate regression procedure. Internal or external calibration based on a single measurement does not cover the mentioned aspects, and contributes additionally to the overall error of the results by transfering the unavoidable random error of the single standard injection to the results calculated therefrom. It is thus suggested that external multiple point calibration in preference to one point calibrations should be used.

Standard injections. It is not an acceptable procedure to inject a couple of standard solutions in the morning and to repeat it after a long series of sample injections. The two standard series would most probably differ considerably; this is especially true in GLC. The detector sensitivity is strongly influenced by changing conditions; within a series of pure standard solution, the detector therefore exhibits a sensitivity different from that within a series of samples due to effects of coextracts. The detector is thus most stable if a steady state is reached, i.e. if standard solutions and samples are injected alternatively (sample-standard-sample-standard etc). The most diluted standard solution must correspond to the desired lower limit of quantitation (cf. Sect. 4.2.6.), the next ones being more concentrated by factors of about 2–3 each. A total of at least 4 different concentrations should be used. The most concentrated standard solution must not exceed the most diluted one by more than a factor of 30 without risk of carry-over effects.

Regression procedure. The following three assumptions have proved valid during many years of experience: (1) detector responses are normally distributed, (2) the standard deviation is proportional to the concentration ("relative error"), (3) there is usually a linear relation between concentration and response. A linear function (Eq. (2)) is therefore taken as the basis for the linear regression ("method of least squares"). The relative errors must be considered for the linear regression, otherwise the dominating deviations at the upper end of the standard

$$y = F(x) = ax + b \quad y = \text{response}, \quad x = \text{injected amount of analyte} \tag{2}$$

function often lead to completely false results at the lower end. Transformation of both response and amount into logarithms is in widespread use to achieve this. This log-transformation is not encouraged for two reasons: it leads to a – although small – systematic error, and the errors are no longer normally distributed, their use to assess the variance of the chromatographic system would not be proper. It should only be selected for the ease of calculation if no computer is available, but the analyst has to be aware of the mentioned deficiencies. The use of the approach for the linear regression as given in Eq. (3) is encouraged if a programmable calculator or a computer is available. It uses the reciprocal standard amounts as weighing factors thus accounting for the amount-dependent nature of errors. This

$$S = \sum_{i=1}^{N} \left(\frac{y_i - F(x_i)}{x_i}\right)^2 \tag{3}$$

x_i = injected amount of analyte
y_i = measured response
N = number of standard injections
S = sum of squares

algorithm is statistically sound without much additional computational effort. The minimum sum of squares is obtained by setting the partial derivatives equal to zero (Eq. (4)). Standard mathematical procedures lead to the parameters a and b.

$$\left[\frac{\delta S}{\delta a}\right]_b = \left[\frac{\delta S}{\delta b}\right]_a = 0 \tag{4}$$

(It is hereby assumed that the analyses are carried out in series of between 10 and 20 samples and the same number of standards). If the injections are done continuously the run would have to be broken into suitable segments.

Variance of the chromatographic system. The correlation coefficient is often used as a measure of the goodness of fit. To quantitatively describe the variance and hence the standard deviation of the calibration it is not suited [25, 26]. The relative deviations of the measured responses from the calculated ones (Eq. (5)) are a better basis for an estimate of the variance. Using the described correlation procedure the mean deviations are zero. The relative variance (Eq. (6)) can be taken as an estimate of the variance of the chromatographic system. The square root of the variance is the relative standard deviation.

$$D_i = \frac{y_i - F(x_i)}{F(x_i)} \qquad D_i = \text{relative difference of calculated and measured response} \tag{5}$$

$$S^2 = \frac{1}{N-1}\sum_{i=1}^{N} D_i^2 \qquad S^2 = \text{variance}, \quad S = \text{standard deviation} \tag{6}$$

The valid working range. Like other mathematical functions the calibration function has a certain range of definition. This range is limited by the lowest and the highest standard amount injected. Extrapolation beyond the range is not valid. This is of utmost importance at the lower end of the function.

Acceptance criteria. The standard deviation of the chromatographic system estimated as outlined above, must generally not exceed 10%. This value reflects stable, well-working systems. No outliers among the standards must be rejected as long as no explanation can be given. Outliers are the result of malfunction or gross errors and would occur not only with standard injection but also with samples, the results would thus be questionable. In case of standard injections containing outliers the whole series has to be re-injected after remedial actions have been taken. The retention times of all standard peaks of an analyte must not vary for more than a few seconds. If they do, appropriate trouble shooting is indicated.

4.2.5 From the Peak to the Result

Once a peak is properly identified and measured the calculation of the result using the inverse calibration function does not cause any difficulties. The main problems in this area are connected with peak identification and measurement.

Peak identification. A peak with its maximum within the retention time window defined by the shortest and the longest retention time among all standard and recovery injections can be considered identified. A peak outside this identification window must be considered not identical with the analyte, i.e. being an interference. If it is tailing into the identification window it could hide a smaller analyte peak, a confirmation analysis using a different chromatographic system is necessary. Otherwise, the result could be falsely negative.

Peak measurement. The main difficulty here is how to draw the baseline. This is critical and needs experience. It should not be left to the computer as is stressed in Sect. 4.2.8. A point of discussion is the question whether peak heights or areas should be measured [27]. In trace analysis peak height is usually better suited, simply because of the often uncertain baseline. Errors in the baseline do much less contribute to peak height than to peak areas. With capillary chromatography peak heights are not always reproducible, peak areas may be preferred in this case.

Reporting. Results have to be reported as found without any deduction of a blank matrix value and without correction for recoveries (it is assumed that these are sufficiently acceptable, see Sect. 4.2.7.). Should, for good reasons, any corrective conversions have to be made, they must remain transparent. Supporting data to accuracy such as comparable recovery values have to be included. Results should never be presented to more digits than statistically reasonable. In trace analysis two significant digits are sufficient, the third one is nothing more than a random number. If it contains any information it is about the analyst's understanding of statistics.

4.2.6 The Lower Limit of Quantitation and Other Limits

"Zero" may be the outcome of a count but it never results from a measurement. Every analyst knows this fundamental truth. The lower end of the measuring scale is called limit of determination or limit of detection, depending on the statistical significance. In the case of chromatographic signals, it is defined as that peak which can reproducibly be differentiated from the noise of a analyte-free check sample at a certain significance level. The determination of this ultimate limit of determination involves multiple measurements. A given method has not a single well-defined limit of determination. It depends on the instrument parameters (which may vary) and on the sample, to mention only the most important factors. Therefore, routine measurements at the statistically defined ultimate limit of determination would require permanently repeated evaluation of that limit under actual conditions. Usually it is neither necessary nor worth the effort to measure routinely at the limit of determination.

A more practical approach is proposed for routine analytical purposes. The lowest concentration to be quantitated is defined according to the actual needs. The method is developed or adapted to be usable in this range. The lowest standard concentration for calibration is selected according to the lowest concentration to be determined and is considered the lower limit of quantitation (LLQ). Of course, the signal of the lowest standard injection has to be well above the theoretical limit of determination, i.e. the signal has to be at least five times as high as the amplitude of the noise in analyte-free injections and the retention time window of interest must be free of interferences. Besides this limitations the LLQ is not related to any blank matrix value.

The LLQ is the mean value of a distribution with a certain standard deviation. Assumed the mean peak height corresponding to LLQ is 1 cm and a peak of a sample was found to be 0.9 cm: Is this result below LLQ? If the measurement is repeated, it might be found to be 1.2 cm, or 0.7 cm, according to the standard deviation of the chromatographic system of ± 10% as agreed upon above; if the whole analysis is repeated the peak could even be found to be 0.5 cm or 1.5 cm. A true value below the LLQ can thus easily result in a measured value of above LLQ and vice versa, illustrating how ambiguous results near limit values can be.

This problem of decision exists at any limit value, not only at the limit of quantitation, even more so at regulatory limits. It is usually of no relevance whether a pesticide concentration is 0.35 or 0.45 µg/1, but it can be important whether it is above or below a limit value, such as a Maximum Admissible Concentration (MAC). If a result close to the limit value is found and the confidence range includes values below the limit or even below zero: How should be decided whether the limit value was exceeded? In case of single results the analyst has the option to take a lower risk of false negatives or to take a lower risk of false positives. The first option should be chosen if evidence has to be given for the absence of an analyte, the second one if the presence of an analyte has to be proved. In the first case all peaks possibly belonging to the distribution of the LLQ (or another limit value) are included, in the other case they are excluded. Taking the "basic variability" as 95% confidence limit, only single values of x+50% are significantly above the limit value x, and only x−50% are significantly below that limit. This does not mean that the actual limit value is increased or decreased, it does show a grey area around limit values where findings cannot be interpreted unambiguously. By analyzing replicate samples and calculating the mean value, the confidence limit becomes narrower as far as random errors are concerned. Systematic errors and gross errors, however, are not averaged out by replicate results.

4.2.7 Recovery Experiments

Quantitation of traces of pesticides in chromatographic systems, as outlined above, implies that the detector response is the same for pure standards and for real samples. This assumption has to be tested by recording a "recovery function" of spiked samples over the whole working range. The variance being roughly the same over the whole range it is usually sufficient to select one spiking level near the lower and one near the upper end of the working range. The recovery values give at the same time show that the standard function is valid to quantitate residues and that the analytical method – and the analyst – were working properly. They represent a method calibration.

Valid recovery measurements have to be carried out at the same time as the analyses of the test samples. It is recommended that, with each series of samples, recoveries be carried out at or – for statistical reasons – slightly above (e.g. by a factor of 2) the lower end of the working range and at or slightly below the upper end. If the lower spiking level exactly matches the lower end of the working range, i.e. is at the LLQ, the analyst is facing problems of quantitation at limit values as outlined in the last paragraph; there are fundamental difficulties to exactly quantitate possible recoveries of less than 100% (which then would be below the LLQ). For recovery experiments water samples of a composition as similar to the real samples but analyte-free are spiked before analysis with the required amount of every analyte. Care has to be taken that the solvent used to dissolve the standard solution is water miscible and has no influence on the analysis.

Acceptable range. The variability of the recovery values is higher than the variability of the chromatographic system as outlined in Sect. 4.2.4. As discussed in Sect. 2.3. the intra-laboratory standard deviation is about ± 15%. If 95% of the statistically possible values are to be included and if a mean recovery of 90% is assumed, the acceptable range for individual recovery values lies between 60% and 120%. If the average recovery of the method is reproducibly lower a ± 30%-limit around that method-typical recovery percentage is acceptable and a correction of the results has to be considered.

Corrections for recoveries. The acceptable range was defined assuming random errors of ± 15%. This means that the results vary within these limits without a reproducible reason. Therefore, using a single recovery value for correcting results would not add to accuracy and is thus discouraged. Only in case the method yields reproducibly (within the frame of ± 15% s.d.) low recoveries, the use of a correction factor may be indicated. The factor should be based on long-term average values in order to correct for systematic method-related errors, but not on randomly varying single results.

4.2.8 The Use of Computers

Computers are a useful tool to facilitate calculation and to handle large numbers of data. As such, they are in widespread use in most analytical laboratories mainly to control instruments including automatic analysis systems, to record and store the output signal of detectors, to integrate chromatograms, to identify peaks, to evaluate peak data, and to calculate calibration functions and results. *In no case should the computer replace critical and scientific judgement by the analyst.*

SOPs may contain the obligation to check the calculation software on each alteration or modification of programs. It is rather easy to get computer programs which are error-free with respect to calculations. Such programs work reliably. This is not the case for those steps involving some type of interpretation or scientific judgement such as peak integration. The core of the integrator program, the algorithm to decide where to draw the chromatographic baseline, where the peaks start, and where they end, is very complicated. This program can be shown to be basically error-free using some typical examples, still this does not mean that the program would adequately integrate each chromatogram. A test is only valid for the chromatogram under consideration. This problem cannot be managed by GLP. There is no other way for the analyst than to check each individual chromatogram for correct peak integration. To this end the output of the chromatographic systems must be a chromatogram, not a compilation of pure figures only, even if the computer collects the data on-line. The analyst needs an analog signal to judge the peak integration.

4.3 Documentation

The analyst has not only to generate accurate results, he must also be able to prove even after years that the results are correct. All results as well as the conclusions based hereon must remain traceable to the original raw data. The documentation must contain sufficient details of the analysis as to fully regenerate the result without need to fill data gaps with assumptions. This includes (1) description of the methodology mentioning all relevant details, (2) calculation procedures, (3) raw data (properly labelled chromatograms, weights, aliquots), (3) information about calibration (characterization of the analytes, weights, volumes, concentrations).

5 Does the Result Need Confirmation?

Confirmation mainly refers to the chemical identity rather than to the figure of the analytical result. In the "National Pesticide Survey" [9] mentioned in Sect. 2.2, the samples giving positive results by properly validated methods were reanalyzed for confirmation on a second column; it was then attempted to qualitatively reconfirm these results once more by GCMS. As much as 45, 55, and 79%, respectively, of the positive results found with three GLC-methods could not be confirmed on the second column, thus turning out to be false positives. Of those confirmed by the second column, another 20% could not be re-confirmed by GCMS. None of the positive findings by the HPLC-method could be confirmed on a second column; all of them were thus false positives. Without proper confirmative analyses, numerous false positives would have been reported in the survey. This example illustrates how important it is that positive findings are confirmed. Non-confirmation of results must thus be considered a gross error.

Falsely identified substances is a perpetual concern in organic trace analyses: Editorials in 1968 [28] and in 1990 [29] reminded authors of the need for confirmation. GLP does not exempt confirmation, on the contrary, not the letter but the spirit of GLP requires confirmation. This does not mean that *each* result must be confirmed, there can be other evidence of the identity of the analyte such as confirmed results in comparable samples. It does mean that a self-critical attitude towards the results requires confirmation if any doubts remain.

5.1 Critical Questions

The critical examination of the results is of utmost importance. The analyst must scrutinize each result for potential irregularities asking the following questions:

- Does the sample exhibit a "normal" chromatographic pattern? I.e. are peaks of solvent and co-extractives present?

- Is the shape of the "identified" peaks symmetric? Is it thus unlikely that the peaks hide other components?
- Are the retention times of the "identified" peaks bracketed by the shortest and the longest retention times among standard and recovery injections?
- Is the result confirmed by other comparable results? Is it "reasonable" in view of the sample history?
- Are the check samples free of peaks in the chromatographic region of interest?
- Can carry-over effects be excluded?

If any of these questions have to be answered with "No" for a given result, a positive one has to be confirmed by a second, independent method, a negative one by careful repetition of the analysis including at least two recovery samples.

5.2 Some Remarks on Confirmation Techniques

Confirmation of positive findings requires methods of a sufficiently different, possibly higher, selectivity. This is meant by "independent" methods. The analyst can make use of selectivities in cleanup, in chromatographic separation technique, or in the detector type. In any case, when interpreting the results of the confirmation study the analyst has to bear in mind that (1) negative outcome of the confirmation clearly proves the absence of the analyte, acceptable recoveries provided, whereas (2) positive outcome means either confirmation or still insufficient selectivity, the critical questions must thus be asked once again. In a recent publication [30] problems of confirmation and examples of insufficient confirmation using different chromatographic columns and detectors are discussed.

Cleanup. Additional cleanup should be of a selective type making use of specific analyte properties. Useful examples are ion exchange columns, adsorption or partition chromatography.

Chromatographic columns. Confirmation columns must be of different type than the previous one. In HPLC a RP-2 instead of a RP-18 would not essentially alter the selectivity. The analyst must aim at using different separation mechanism. Best results are obtained with multidimensional HPLC, i.e. column switching.

Detectors. The method of choice is mass spectrometric detectors, the GLC-type is already in widespread use, as yet less the HPLC-type. Even these detectors can fail in confirmation if used on only a single mass [31]. GCMS does thus not generally dispense with critical examination of the result. When using a multi-wavelength detector in HPLC, the selected wavelengths must be sufficiently distinct and the results must be quantitatively identical, because the UV-bands are broad and thus do not offer sufficient selectivities in all cases.

The best confirmation technique is to combine a completely different method with selective detectors: Confirmation of GLC-results with a HPLC-method and vice versa. The confirmation technique used must have a sufficiently low limit of determination as to allow a quantitative confirmation and the same precautionary measures must be observed as with the previous analysis.

6 Conclusions

With the methodology available today it is possible to perform pesticide determinations at the sub-μg/l level with an inter-laboratory "basic variability" of ± 50% (95% confidence limits), provided that an appropriate laboratory-internal working system to assure data quality is established. This working system must make sure that

- valid methods are used by well-trained and science-minded analysts,
- gross errors are avoided and accuracy is controlled by adequate measures focussing on the most sensitive parameters of the analytical methods,
- results are critically reviewed and confirmed,
- an accuracy evaluation is included, and
- a comprehensive documentation provides complete data-tracking.

GLP is only suited to meet these quality requirements if the relevant guidelines are interpreted in a science-oriented way. Otherwise it would only cover the formal aspects.

The quality of the data is an essential prerequisite for the quality of conclusions based on them. In addition to the analytical results, all the pertinent information must be taken into consideration. Sufficient information must be provided to allow conclusions to be verified.

7 References

1. Oppenhuizen ME, Cowell JE (1991) J. Assoc. Off. Anal. Chem. 74: 317
2. Edgell KW, Biederman LA, Longbottom JE (1991) J. Assoc. Off. Anal. Chem. 74: 309
3. Edgell KW, Jenkins EL, Lopez-Avila V, Longbottom JE (1991) J. Assoc. Off. Anal. Chem. 74: 295
4. Cowell JE, Kunstman JL, Nord PJ, Steinmetz JR, Wilson GR (1986) J. Agric. Food Chem. 34: 955
5. Lopez-Avila V, Wesselman R, Edgell K (1990) J. Assoc. Off. Anal. Chem. 73: 276
6. Alford-Stevens AL, Eichelberger JW, Budde WL (1988) Environ. Sci. Technol. 22: 304
7. Legler J (1991) Fresenius' J. Anal. Chem. 339: 417
8. Munch DJ, Graves RL, Maxey RA, Engel TM (1990) Environ. Sci. Technol. 24: 1446
9. U.S. Environmental Protection Agency Washington (1990) National Survey of Pesticides in Drinking Water Wells Phase I Report EPA 570/9-90-15
10. Tashiro C, Clement RE, Davies S, Oliver B, Munshaw T, Fenwick J, Chittim B, Foster MG (1990) Chemosphere 20: 1313
11. Horwitz W, Kamps LR, Boyer KW (1980) J. Assoc. Off. Anal. Chem. 63: 1344
12. Benoit FM, LeBel GL (1986) Bull. Environ. Contam. Toxicol. 37: 686
13. Norris LA (1986) Weed Sci. 34: 485
14. Guth JA (1991) Gewässerschutz, Wasser, Abwasser 71: 835

15. Weil L, Hörmann WD (1991) Umweltwiss. Schadst. Forsch. 3: 306
16. Horwitz W (1982) J. Assoc. Off. Anal. Chem. 65: 525
17. OECD Paris (1981) Guidelines for Testing Chemicals, "Principles of Good Laboratory Practices"
18. Merz W, Weberruß U, Wittlinger R (1992) Fresenius' J. Anal. Chem. 342: 779
19. GIFAP Brussels (1990) Water Quality Monitoring: Site Selection and Sampling Procedures for Pesticide Analysis Technical Monograph No. 13
20. Keith LH (1992) Environmental Sampling and Analysis: A Practical Guide Lewis Publishers, Chelsea MI (USA)
21. House WA, Ou Z (1992) Chemosphere 24: 819
22. Munch DJ, Frebis CP (1992) Environ. Sci. Technol. 26: 921
23. Cresser M (1990) Anal. Proc.(London) 27: 110
24. Foreman WT, Zaugg SD, Faires LM, Werner MG, Leiker TJ, Rogerson PF (1992) Environ. Sci. Technol. 26: 1307
25. Colburn WA, Gibaldi M (1978) Can. J. Pharm. Sci. 13: 31
26. VanArendonk MD, Skogerboe RK, Grant CL (1981) Anal. Chem. 53: 2350
27. Kipiniak W (1981) J. Chromatogr. Sci. 19: 332
28. Schlechter MS (1968) Pestic. Monit. J. 2: 1
29. Glaze WH (1990) Environ. Sci. Technol. 24: 271
30. Koskinen WC, Otto JM, Jarvis LJ, Dowdy RH (1992) J. Environ. Sci. Health B27: 255
31. Benfenati E, Tremolada P, Chiapetta L, Frassanito R, Bassi G, Di Toro N, Fanelli R, Stella G (1990) Chemosphere 21: 1411

Subject Index

Previously Published Volumes in Chemistry of Plant Protection

Vol. 1

Sterol Biosynthesis, Inhibitors and Anti-Feeding Compounds

T. Kato
Sterol-Biosynthesis in Fungi, a Target for Broad Spectrum Fungicides

D. W. Krämer
Chemistry of Sterol Biosynthesis Inhibiting Fungicides

K.-H. Kuck, H. Scheinpflug
Biology of Sterol-Biosynthesis Inhibiting Fungicides

D. M. Norris
Anti-Feeding Compounds

Vol. 2

Degradation of Pesticides, Desiccation and Defoliation, ACh-Receptors as Targets

P. R. Wallnöfer, G. Engelhardt
Microbial Degradation of Pesticides

U. Schoknecht, D. Otto
Enzymes Involved in the Metabolism of Organophosphorus, Carbamate and Pyrethroid Insecticides

R. Richter, D. Otto, H.-J. Mengs
Insecticide Compounds, Acting on the Acetylcholine Receptor of the Insect Nervous System

H. Bermann, D. Martin
Chemical Manipulation of Desiccation and Defoliation and Essential Aspects for the Application and Development of New Chemical Compounds in the Future

Vol. 3

Pyrethroid Residues, Immunoassays for Low Molecular Weight Compounds

W. Blaß
Residues of Pyrethroid Insecticides and Their Analytical Determination

E. W. Weiler
Immunoassay for the Quantitation of Low Molecular Weight Compounds in Plants

Vol. 4

K. Naumann
Synthetic Pyrethroid Insecticides: Structures and Properties

Vol. 5

K. Naumann
Synthetic Pyrethroid Insecticides: Chemistry and Patents

Vol. 6

Controlled Release, Biochemical Effects of Pesticides, Inhibition of Plant Pathogenic Fungi

M. Bahadir, G. Pfister
Controlled Release Formulations of Pesticides

M. Eto
Biochemical Mechanisms of Insecticidal Activities

M. A. Q. Khan
Biochemical Effects of Pesticides on Mammals

P. Böger, G. Sandmann
Modern Herbicides Affecting Typical Plant Processes

H. Buchenauer
Physiological Reactions in the Inhibition of Plant Pathogenic Fungi

Vol. 7
Herbicide Resistance - Brassinosteroids, Gibberellins, Plant Growth Regulators

D. Gross
Plant Growth Regulatory Substances Both of Microbial and Plant Origin

G. Adam, B. Voigt, M. Lischewski
Synthesis and Labelling of Gibberellins

V. Marquardt, G. Adam
Recent Advances in Brassinosteroid Research

K. C. Vaughn, S. O. Duke
Biochemical Basis of Herbicide Resistance

Vol. 8
Terrestrial Behavior of Pesticides

I. Scheunert
Physical and Physico-Chemical Processes Governing the Residue Behavior of Pesticides in Terrestrial Ecosystems

I. Scheunert
Transformation and Degradation of Pesticides in Soil

I. Scheunert
Fate of Pesticides in Plants and in Soil Fauna

H. Parlar
Mechanism for the Behavior of Pesticides on Surfaces

Vol. 9

Pesticides in Ground and Surface Water

H. Beitz, H. Schmidt, F. Herzel
Occurrence, Toxicological and Ecotoxicological Significance of Pesticides in Groundwater and Surface Water

D. W. Bewick
The Mobility of Pesticides in Soil - Studies to Prevent Groundwater Contamination

C. N. Guyot
Strategies to Minimize the Pollution of Water by Pesticides - A Critical Review to Plant Protection Measures in the Past, Present and Future

M. O. James
Pesticide Metabolism in Aquatic Organisms

G. Mattheß
Fate of Pesticides in Aquatic Environments

M. Häfner
Pesticides in Soil: A German Approach of Predicting Their Movement into Ground and Surface Water

Vol. 10

Herbicides Inhibiting Branched - Chain Amino Acid Biosynthesis - Recent Developments

J. Stetter
Introduction

J. V. Schloss
Recent Advances in Understanding the Mechanism and Inhibition of Acetolactate Synthase

S. McGee, J. V. Hay
Recent Developments in the Chemistry of Sulfonylurea Herbicides

H. M. Brown, J. C. Cotterman
Recent Advances in Sulfonylurea Herbicides

D. W. Ladner
Recent Studies of Imidazolinone Herbicides and Related Compuonds

W. A. Kleschick
Triazolopyrimidine Sulfonanilide Herbicides and Related Compounds

B. C. Gerwick, G. J. DeBoer, P. R. Schmitzer
Mechanisms of Tolerance to Triazolopyrimidine Sulfonanilide Herbicides

M. W. Drewes
Salicylic Acids and Related Structures

F. Lieb, U. C. Philipp
Miscellaneous Structures

Printing: Saladruck, Berlin
Binding: Buchbinderei Lüderitz & Bauer, Berlin